Mastering CSS

Cascading Style Sheets (CSS) is an open-source programming language used in website building and HTML templates that integrates all relevant information related to web page displays. CSS is used to format the look and structure of a web page, as well as to set design features such as basic layout, colors, and fonts. CSS allows for continuity between different web pages on the website and makes webpage development easier and faster.

This book has been created to help readers understand and learn the concepts of CSS. It discusses the fundamental concepts of CSS, including its properties and functions, and guides the reader through creating websites with it.

Key Features:

- Examines the fundamentals of CSS, values, selectors, and queries

- Discusses its application in modern web development to help readers to quickly advance the necessary information

- Explores animations, grids, flexboxes, masking, filtering, and compositing using CSS

Mastering CSS is a valuable resource for anyone who wants to create a website. After finishing this book, readers will quickly build their website with absolute ease, even if they were utterly oblivious to it before.

About the Series

The *Mastering Computer Science* covers a wide range of topics, spanning programming languages as well as modern-day technologies and frameworks. The series has a special focus on beginner-level content, and is presented in an easy-to-understand manner, comprising:

- Crystal-clear text, spanning various topics sorted by relevance,

- Special focus on practical exercises, with numerous code samples and programs,

- A guided approach to programming, with step-by-step tutorials for the absolute beginners,

- Keen emphasis on real-world utility of skills, thereby cutting the redundant and seldom-used concepts and focusing instead of industry-prevalent coding paradigm,

- A wide range of references and resources, to help both beginner and intermediate-level developers gain the most out of the books.

Mastering Computer Science series of books start from the core concepts, and then quickly move on to industry-standard coding practices, to help learners gain efficient and crucial skills in as little time as possible. The books assume no prior knowledge of coding, so even the absolute newbie coders can benefit from this series.

Mastering Computer Science series is edited by Sufyan bin Uzayr, a writer and educator with over a decade of experience in the computing field.

For more information about this series, please visit: https://www.routledge.com/Mastering-Computer-Science/book-series/MCS

Mastering CSS
A Beginner's Guide

Edited by
Sufyan bin Uzayr

CRC Press
Taylor & Francis Group
Boca Raton London New York

CRC Press is an imprint of the
Taylor & Francis Group, an **informa** business

First Edition published 2024
by CRC Press
2385 NW Executive Center Drive, Suite 320, Boca Raton, FL 33431

and by CRC Press
2 Park Square, Milton Park, Abingdon, Oxon, OX14 4RN

CRC Press is an imprint of Taylor & Francis Group, LLC

© 2024 Sufyan bin Uzayr

Library of Congress Cataloging-in-Publication Data

Names: Bin Uzayr, Sufyan, editor.
Title: Mastering CSS : a beginner's guide / edited by Sufyan bin Uzayr.
Description: First edition. | Boca Raton : CRC Press, 2024. | Series:
    Mastering computer science | Includes bibliographical references and
    index.
Identifiers: LCCN 2023004868 (print) | LCCN 2023004869 (ebook) | ISBN
    9781032414355 (hardback) | ISBN 9781032414324 (paperback) | ISBN
    9781003358060 (ebook)
Subjects: LCSH: Cascading style sheets. | Web sites--Design. | Web site
    development.
Classification: LCC TK5105.888 .M3656 2024 (print) | LCC TK5105.888
    (ebook) | DDC 006.7/4--dc23/eng/20230406
LC record available at https://lccn.loc.gov/2023004868
LC ebook record available at https://lccn.loc.gov/2023004869

ISBN: 9781032414355 (hbk)
ISBN: 9781032414324 (pbk)
ISBN: 9781003358060 (ebk)

DOI: 10.1201/ 9781003358060

Typeset in Minion
by KnowledgeWorks Global Ltd.

For Mom

Contents

About the Editor

Sufyan bin Uzayr is a writer, coder, and entrepreneur having over a decade of experience in the industry. He has authored several books in the past, pertaining to a diverse range of topics, ranging from History to Computers/IT.

Sufyan is the Director of Parakozm, a multinational IT company specializing in EdTech solutions. He also runs Zeba Academy, an online learning and teaching vertical with a focus on STEM fields.

Sufyan specializes in a wide variety of technologies such as JavaScript, Dart, WordPress, Drupal, Linux, and Python. He holds multiple degrees, including ones in Management, IT, Literature, and Political Science.

Sufyan is a digital nomad, dividing his time between four countries. He has lived and taught in universities and educational institutions around the globe. Sufyan takes a keen interest in technology, politics, literature, history, and sports, and in his spare time, he enjoys teaching coding and English to young students.

Learn more at sufyanism.com

Acknowledgments

There are many people who deserve being on this page because this book would not have come into existence without their support. That said, some names deserve a special mention, and I am genuinely grateful to:

- My parents, for everything they have done for me.

- The Parakozm team, especially Divya Sachdeva, Jaskiran Kaur, and Simran Rao, for offering great amounts of help and assistance during the book-writing process.

- The CRC team, especially Sean Connelly and Danielle Zarfati, for ensuring that the book's content, layout, formatting, and everything else remain perfect throughout.

- Reviewers of this book, for going through the manuscript and providing their insight and feedback.

- Typesetters, cover designers, printers, and everyone else, for their part in the development of this book.

- All the folks associated with Zeba Academy, either directly or indirectly, for their help and support.

- The programming community in general, and the web development community in particular, for all their hard work and efforts.

Sufyan bin Uzayr

Zeba Academy – Mastering Computer Science

The "Mastering Computer Science" series of books are authored by the Zeba Academy team members, led by Sufyan bin Uzayr, consisting of:

- Divya Sachdeva

- Jaskiran Kaur

- Simran Rao

- Aruqqa Khateib

- Suleymen Fez

- Ibbi Yasmin

- Alexander Izbassar

Zeba Academy is an EdTech venture that develops courses and content for learners primarily in STEM fields, and offers educational consulting and mentorship to learners and educators worldwide.

Additionally, Zeba Academy is actively engaged in running IT Schools in the CIS countries, and is currently working in partnership with numerous universities and institutions.

For more info, please visit https://zeba.academy

CHAPTER 1

CSS

Introduction

IN THIS CHAPTER

- ➤ Introduction
- ➤ History
- ➤ Version of CSS
- ➤ Basic HTML
- ➤ Types of CSS
- ➤ Various Properties (Classes, Id, Divisions)

In this chapter, we will learn about CSS (Cascading Style Sheets), its fundamental concepts, versions, syntax, and how we can link our HTML page with CSS files. The other concepts like modules, selectors, ids, borders, attributes, tables, form, colors, etc. will also be covered in this whole lesson. So let's get started with the basics of CSS.

INTRODUCTION

A document is usually a text file created using tag language – HTML is the common markup language, but you can also merge with other tag languages such as SVG or XML. It is presenting a document to other user means converting it into a form used by your audience. The other browsers,

DOI: 10.1201/9781003358060-1

1

such as Firefox, Chrome, or Edge, are designed to deliver text by a view, for example, on a computer screen, projector, or printer.

Cascading Style Sheets (CSS) is a programming language that integrates all relevant information related to web page displays. CSS defines the style and format of a website or page, including layout, colors, fonts, padding (space around each element), and more. Along with HTML and Javascript, CSS forms the basis of how the Internet works. All three standards and specifications are also maintained by the World Wide Web Consortium (W3C).

CSS can be used to create a very basic text style – for example, to change the color and size of titles and links. It can be used to build a structure, for example, to convert a single column of text into a structure with the main content area and a different bar for related information. It can be used for effects such as animation.

HISTORY

CSS is highly regarded by the Norwegian Håkon Wium Lie, who in 1994 wanted to create a standard style sheet for the World Wide Web. The first website Lie tried for CSS is the Arena web browser. Since its initial creation, Lie has gone on to co-produce versions of CSS-1, CS-S2, and RFC 2318 with Tim Berners-Lee and Robert Cailliau. In its first decade of existence (1994–2004), CSS in all its clarity became a web standard that had a significant impact on the look of the World Wide Web as we know it now. CSS3 was released in 1999.

CSS is one of the three core technologies used on the web (the other two being HTML and JavaScript). CSS stands for Cascading Style Spreadsheets – the links are actually in terms "cascading" and "style" with cascading describing how one style can go from one to another.

One of the various benefits of CSS is that more than a single style can be used within a single HTML document. CSS is used to define what HTML code will look like on a website. Although HTML (Hypertext Markup Language) is used to create content, including text, CSS changes the way a web page will look. Thus, depending on the data they want to display, the developer may choose to have a page with other tabs running at the top of the page or on the side.

Alternatively, some developers may choose to use titles and subtitle styles to ensure that words appear on the page or alter or redesign the existing web page. Perhaps the best way to describe what CSS does is to explain that the page would like to be free to use CSS. Without CSS web pages are clear and far from inspiring. Words scroll through the page and read hard. But, before CSS that was exactly what web pages looked like. The introduction of CSS is responsible in part for what the web looks like and feels today. And, far from being created and as a result, it is a continuously evolving language. Web standards are a topic close to Lie's heart. Since the introduction of CSS, he has appealed to major technology players like Microsoft and other browsers to support the standard web standards and continue to improve.

WHY DO WE NEED CSS?

First, applying CSS ensures that your web pages are consistent. If your website has 100 pages, now imagine that you have to enter a code to define title sizes, layout, and other display data and combine all the content each time you want to produce a new web page. Also, imagine having a 100-page site and being able to change one of them while keeping everything the same.

CSS also makes that possible. Using CSS brings flexibility where needed, but it is flexible enough to be able to make changes to an individual page or a section.

WHY DO WE USE CSS?

For a website to work properly, it must have a fast download time. Nowadays, people usually wait a few seconds for a website to load. Therefore, it is important to ensure a fast pace. For companies looking to ensure fast and smooth website information, CSS becomes key to their success.

CSS is easy to maintain due to the short maintenance time. This is because single-line code conversion affects the entire web page. Also, if upgrades are needed, make a small effort.

You would not find many good and easy-to-use websites. One thing that is common to all of these websites is the consistency of construction. CSS empowers developers to ensure that style features are applied consistently across every few web pages.

Due to its fast speed and easy maintenance, CSS saves a lot of time and effort in the web development process due to its fast loading time. Here, a little time ensures the good performance of the designer.

People use different smart devices to view a particular website. It can be a smartphone, PC, or laptop. For this purpose, websites are required to be compatible with the device. CSS ensures smooth operation by providing better alignment.

You can change the location of the HTML tag with the help of CSS. You can set things as an image on any part of a web page as needed.

DIFFERENT VERSION OF CSS

- **CSS1 and CSS2 (no longer applicable)** – Released from W3C as a recommendation in December 1996. This version defines the CSS language and the visual formatting model for all HTML tags. These were the two versions of CSS that are no longer updated or maintained.

- **CSS2.1 (recommended)** – W3C recommendation in May 1998 and built on CSS1. This version adds support for specific media style sheets, for example, audio printers and devices, downloadable fonts, object layouts, and tables. This version has fixed many bugs and problems in CSS2 and is now the official, recommended version of CSS.

- **CSS3** – CSS3 became a W3C recommendation in June 1999 and was built on older versions of CSS. It is divided into texts called Modules, and here each module has new extension features defined in CSS2. These versions build in CSS2.1, adding new functionality and maintaining background compliance. Some features are still being tested and may change in the future. Use this with caution, as it may cause problems with your site.

  CSS3 is the latest standard for CSS earlier versions (CSS2). The main differences between CSS2 and CSS3 are as follows:

  - Media Queries

  - Namespaces

  - Selectors Level 3

  - Color

- CSS4 has never become an official version.

MAJOR DIFFERENCES BETWEEN CSS, CSS2, AND CSS3

CSS was originally released in 1996 and contains features to add font layouts such as typeface and text color emphasis, background, and other features. CSS2 was released in 1998 with additional styles for other media types to be used to design page layouts. CSS3 was released in 1999 and presentation style structures were added to it, allowing you to create a presentation from documents.

With CSS3, designers can now use special fonts, such as those found on Google Fonts and Typecast. Previously, with CSS and CSS2, designers could only use "secure web fonts" to ensure 100% use of fonts that would always show the same throughout the machine.

CSS SPECIFICATION

All web-level technologies (HTML, CSS, JavaScript, etc.) are defined in large documents called specifications (or "specs"), published by standard organizations (such as W3C, WHATWG, ECMA, or Khronos) and describe accurately how those technologies should behave. CSS is no other – it is developed by an internal W3C team called the CSS Working Group. This group is made up of representatives of browser vendors and other companies interested in CSS. There are also other people, known as invited experts, who act as independent voices; affiliated with the member organization.

New CSS features are developed by the CSS Working Group sometimes because a specific browser is interested in a specific skill, sometimes because web designers and developers request a feature, and because the Working Group has identified a requirement. CSS is constantly evolving, with new features available. However, the important thing about CSS is that everyone works hard so that they never change things in a way that can break old websites. A website built in 2000 that used the limited CSS that existed then, should still be used in the browser today!

As a newcomer to CSS, you may find the CSS specification too great – it is intended for developers to apply feature support to user agents, not for web developers to learn to understand CSS. Many experienced developers would prefer to look at MDN documents or other courses. However, it is worth knowing that these features exist and to understand the relationship between the CSS you are using, browser support (see below), and details.

CSS MODULES

As there are many things you can style using CSS, the language is divided into modules. You will see a reference to these modules as user explore MDN. Many text pages are organized around a specific module. For example, you can refer to the MDN reference in the Background and Boundaries module to find out what its purpose is with the properties and features it contains

In this section, you do not need to worry much about how CSS is created; however, it may be easier to get information if, for example, you know that a certain place may be found among other things, therefore, it is almost in the same situation.

For example, let's go back to the Background and Borders module – you may think it makes sense that background color and border color structures are defined in this module. And you will be right. If you know something about HTML, then it is good; if not, let's have a brief introduction to HTML and how to link a CSS file with HTML for changing styles.

BASIC HTML

The following section provides an introduction to HTML. We cannot imagine web pages and the World Wide Web without HTML. It is the language most used to write web pages. It represents Hyper-Text Mark-up Language. You should know that any link found on web pages is usually called Hypertext, and the mark refers to the tag or page layout in such a way that the text in the webpage list is displayed in the correct format. The purpose of developing HTML was to understand the structure of any text: title, body, content, or categories.

So, basically, HTML provides a structured format to display the content of web pages. It is very simple and easy to understand. In the early nineties, it was developed by Tim Berners-Lee and later went through many changes and modifications. HTML 5 is the latest version of Html.

You know that HTML is a simple language that can use many tags to format content. All tags are enclosed within angular brackets <tagname>. With the exception of a few tags, most tags start with angular tags and end with corresponding angular tags.

<! DocType Html> defines the document type and version of Html. The Html code starts just after the <html> angular marker and ends with </html> not seen in the screenshot above.

It usually has two major parts, namely the head and the body. Each category has its set of requirements related to requirements. You will get various sections in HTML structure. Let's discuss each.

Head Section

The header tag represents a web document header that you can tag <head> and <link> with. It starts with <head> and ends with </head>. It has parts of the title inside. For example, <title> This will be your webpage title. </li>>.

Body Section

It represents the body of a web document that usually contains titles, text, and sections. Topics start with <title> and end with </title>. Among these tags, the content can be labeled as "this is the first topic."

The paragraph will start with <p> and end with </p>. The content of the section should be written between these angular characters. The basic Html code shown below in the overview section is used to create a simple Html page.

```
<!DOCTYPE html>
<html lang="en">
<meta charset="utf-8">
<title> Page Title </title>
<body>
    <h1> This is a Heading </h1>
    <p> This is a paragraph </p>
</body>
</html>
```

There are various elements we can add to our HTML code; let's take a look.

- HTML headings are elements defined by tags <h1> to <h6> where <h1> defines the important tag and <h6> defines the less important tag. The HTML section is an HTML element that will be defined using the <p> tag.

- HTML images are a feature of HTML and are defined by the <img> tag, and need to specify attributes such as image src, alt means other text, width, and height.

- HTML lists are elements and are defined using <ul> or <ol> tags where <ul> is a random list and <ol> has an ordered list.

- The HTML table is part of the HTML and can be defined using tag <table> and tag lines <tr> and tag cells <td>.

- HTML links are elements, and can be defined using the <a> tag and sample code below:

  - The HTML attribute style can be used with a combination of any elements such as <p>.

- In HTML, we use the lang attribute, we can say the language of a document using the <html> tag and the language defined using the lang attribute.

- In HTML, we can use the formatting elements to format the document and we can define special text features that have a special meaning. HTML elements like <b> bold, <i> by italics.

- In HTML, we can highlight specific text in a document using the <mark> feature to highlight text included in the <mark> element.

- In HTML, we can define a text as the text above using the <sup> element in the HTML document so that the text embedded in the <sup> element becomes larger text.

The <!DOCTYPE> represents the document type and helps browsers to display web pages correctly. It appears once, at the top of the page (before any of the HTML tags). The <!DOCTYPE> declaration is not case sensitive.

Wherever the web is, it is HTML. HTML usage is distributed across all devices. Here is the list of features of HTML to know where we can use it.

- Browsers like Chrome, Firefox, and Safari all use HTML to render content on the web for better display.

- Various mobile browsers like Opera, Firefox Focus, Microsoft Edge, Dolphin, and Puffin all use HTML to better present and visualize online content on mobile phones.

- Various smart devices are embedded with HTML functions to better browse and navigate during their operations.

- HTML supports the first channel verification method on any web page to stop unwanted traffic.

- HTML accepts great content but gives the same visibility to smaller screen devices and larger screen devices.

- HTML supports a variety of colors, formats, and layouts.

- HTML uses templates that make website design easy.

- The HTML and XML syntax are very similar, so it is easy to work between the two domains.

- FrontPage, Dreamweaver, and many other development tools support HTML.

STARTING WITH A SPECIFIC HTML

HTML and CSS are two methods of tags (code), with their own unique syntax. There is a difference between the two. You can think of HTML as a page layout, while CSS gives HTML its own style.

```
HTML = structure
CSS = style
```

LINKING YOUR HTML AND CSS FILES

Before we can write CSS, we have to go back to HTML. You need to write a new line to link the HTML file and CSS file together. So, open the HTML file and add the provided line "<link href =" style.css "rel =" stylesheet "type =" text / CSS "/>". Your file code looks like this,

```
<!DOCTYPE html>
<html>
 <head>
  <title>This is page title.</title>

  <!-- Here is the External Style Sheet -->
  <link type="text/CSS"  rel="stylesheet"
href="style.css" />

  <!-- Here is the Internal Style  -->
  <style>
      h2, p {
          font-size: 24px
      }
      h1{
          text-align: center;
          font-size: 36px;
          font-family: 'Franklin Gothic Medium',
'Arial Narrow', Arial, sans-serif;
      }
      .container{
          align-items: center;
          max-width: 600px;
          max-height: auto;
          margin: 280px auto;
          border: 1px solid black;
          border-radius: 20px;
```

```
            padding: 50px;
        }
   </style>
   </head>
   <body>
      <div class="container">
         <h1> <u> This is an example of Simple CSS </u>
</h1>
         <!-- Here is the Inline Style -->
         <h2 style="color: red"> This is a heading
element </h2>
         <p style="color: purple"> Hello world, this is
a paragraph. </p>
      </div>
   </body>
<html>
```

This is an example of Simple CSS

This is a heading element

Hello world, this is a paragraph.

Example of simple CSS.

Above line of code links a new CSS file to your HTML file. Let's split it: the href attribute specifies the link associated with the CSS file. We'll get to the links later, for now, just make sure that style.css file is in the same folder as index.html file. The rel attribute tells your browser that this is a style sheet. The attribute type tells the browser that the linked file should be translated as CSS syntax.

HOW CSS AND HTML WORK TOGETHER

CSS is only concerned with web page layout, while page content is defined using marker language such as HTML. The separation of style and content has various advantages, among them improved accessibility, and more control over web design.

CSS documents are used to define a web page style, then linked to an HTML document (or document in a different tag language) that contains the content and layout of the page. Setting the style directly to the HTML document is possible but not recommended. CSS texts can be created in any text editor, such as Text Editor on Mac or Notepad for Windows, as well as many other free also paid options that you can download.

HOW DOES CSS WORK?

CSS contains the rules and values that a web browser can use in the content of a web page to better display its content. For example, you can use to define that the body part of a page has a blue background, that the text is displayed in a white Helvetica font with a size of 18px. CSS rules are read by a specific document format (hence the word "cascading" in Cascading Style Spreadsheets). In general, the next step in a CSS document is the winner – unless the first rule is very clear. For example, you might have two rules for your conflicting CSS file – such as setting body font in blue and section font in green. In this case, the effective rule will be the category font, because it is much clearer than the body font.

CSS SYNTAX

Now move on to the original CSS. The first thing we do is make the paragraph content a different color. So the type or paste into your style file.css.

```
p {color: blue; }
```

This looks different from the code in the HTML file because it is a different syntax. I will add a white area and cut into that code as follows:

```
p {
    color: blue;
}
```

Both of the above examples are exactly the same about your browser. But developers often write CSS as the latest example to differentiate styles.

CSS is a rule-based language – defines rules by specifying groups of styles to be applied to specific objects or groups of objects on your web page. For example, you may decide that a major topic on your page is shown as a red flag. The following code shows a simple CSS rule.

```
p {
    color: red;
    font-size: 5em;
}
```

- For example, the CSS rule opens with a selector.

- Selects the HTML element that will style it. In this case, we style the first-level titles (<p>).

- Then we have a set of twisted pieces {}.

- Inside the instruments, there will be one or more announcements, which take the form of the goods and pairs of values. It specifies the location (color in the example above) before the colon and the value of the property after the colon (red for example).

- This example contains two declarations, one color and one font size. Each pair specifies the location of the element (s) we select (<p> in this case), and then the value we would like to give the structure.

CSS layouts have different valid values, depending on the specified format. In our example, we have a color package, which can take on different color values. We also have a font size structure. This structure can take units of various sizes as a value.

The CSS style sheet will contain many such rules, written in sequence.

```
p {
    color: red;
    font-size: 5em;
}

h2 {
    color: black;
}
```

INHERITANCE CSS

When you nest an element inside another, the nested element inherits the properties assigned to the containing element. Unless you change the values of the inner items independently. For instance, a font declared in the body is inherited by all text in the file, regardless of the item it contains, unless you declare another font for a particular nested item.

Sample:

```
body {font-family: Verdana, serif;}
```

All text in the HTML file will now be set to the Verdana font family. If you want to format certain text with another font like h1 or paragraph, you can do the following.

Sample:

```
p {font-family: Tahoma, serif;}, h1 {font-
family: Georgia, sans-serif;}
```

COMMENT TAGS

Comments can be used to explain why you're adding certain selectors to your CSS file. To help others who might see your file or remember what we thought at a later date. You can add comments that would be ignored by browsers as given below:

/* This is a comment */ starts with / (slash), followed by * (asterisk), followed by the comment, followed by the closing tag immediately after the opening tag * (asterisks) followed by / (slash).

Merge selectors.

You can combine items in a single selector as follows.

Sample:

```
h1, h2, h3, h4, h5, h6
{
    color: #009900;
    font-family: Georgia,
    sans Serif;
}
```

As you can look on the code above, we've grouped all the header elements into a single selector. Each one is separated by a comma. The final result of the above code will set all headers to green and to the specified font. If the user doesn't have the first font we declared, they will go for another sans-serif font.

DIFFERENT TYPES OF CSS YOU CAN USE

There are three ways to write CSS for a web page. You have to decide which one is better for you, but we recommend using an External Style Sheet, which is the third type we'll learn today.

Here are three ways to use CSS:

1. **Inline styles:** It is placed inside HTML elements.

2. **Internal styles:** It is placed in the <head> tag section of the web page you are writing.

3. **External Styles:** It is placed in the External Style Sheet, which is a separate page that links to the web page (recommended).

An explanation of CSS methods is given below:

- **External style sheet:** The external CSS file can be created using any text or HTML editor like "Notepad" or "Dreamweaver." A CSS file contains no HTML elements tags, only CSS. Now save it with the .css file extension. You can link the file externally by keeping that one of the following links in the head section of each (X)HTML file that you want to style with CSS file.

```
<link type="text/css"  rel="stylesheet"
href="Path To stylesheet.css" />
You can also use the @import method like <style
type="text/css">  @import url('Path_to stylesheet.
css')  </style>.
```

Example:
```
<head>
<title> Title <title>
<link  type="text/CSS" rel="stylesheet"
href="style.css" />
</head>
  <body>
```

or,

```
<head>
<title> Title <title>
```

```
<style type="text/css"> @import url('Path of
stylesheet.css') </style>
</head>
<body>
```

Example:

Index.html

```
<!DOCTYPE html>
<html>
 <head>
  <title>This is page title.</title>
  <!-- Here is the Internal Style  -->
  <link href="./styless.css" rel="stylesheet">
 </head>
 <body>
    <div class="container">
        <h1> <u> The HTML for an internal
stylesheet  </u> </h1>
        <h2> This is a heading element </h2>
        <p> This is a paragraph. </p>
    </div>
 </body>
<html>
```

Styless.css

```
h2, p {
    font-size: 32px;
      color: lightseagreen;
      text-shadow: 2px 2px black;
  }
  h1 {
      font-size: 42px;
      font-family: 'Franklin Gothic Medium',
'Arial Narrow', Arial, sans-serif;
  }
  .container {
      align-items: center;
      max-width: 800px;
      max-height: auto;
      margin: 280px auto;
```

```
border: 1px solid black;
border-radius: 20px;
padding: 50px;
}
```

The HTML for an internal stylesheet

This is a heading element

This is a paragraph.

The external style sheet.

- **Internal style sheet:** It is a way you are simply placing that CSS code within the <head> and </head> tags of each HTML file you want to style with the CSS. The format is shown in the example below:

```
<html>
 <head>
<title> Title -  Webpage <title>
<style type="text/CSS">
<body>
     CSS Content Goes Here
</style>
 </head>
<body>
</html>
```

With the above method, each HTML file contains the CSS styling code needed to style the page. That means any changes you want to make to a page, will have to be made to all. The method can be good if you need to style only a page, or if you want other pages to have varying styles.

Example:

```
<!DOCTYPE html>
<html>
 <head>
```

```
<title>This is page title.</title>
<!-- Here is the Internal Style  -->
<style>
    h2, p {
      font-size: 32px;
        color: lightseagreen;
        text-shadow: 2px 2px black;
    }
    h1 {
        font-size: 42px;
        font-family: 'Franklin Gothic Medium',
'Arial Narrow', Arial, sans-serif;
    }
    . container {
        align-items: center;
        max-width: 800px;
        max-height: auto;
        margin: 280px auto;
        border: 1px solid black;
        border-radius: 20px;
        padding: 50px;
    }
 </style>
</head>
<body>
    <div class="container">
        <h1> <u> The HTML for an internal
stylesheet </u> </h1>
        <h2> This is a heading element </h2>
        <p> This is a paragraph. </p>
    </div>
</body>
<html>
```

<u>**The HTML for an internal stylesheet**</u>

This is a heading element

This is a paragraph.

The internal style sheet.

- **Inline styles:** It is a defeat purpose of using CSS in the first place. Inline styles are defined well in the HTML file alongside the element you want to style. It may appear in the queue, next to the CSS text.

```
<div style="background-color: yellow">  This is
inline text </div>
 <p style="color: #ff0000;"> This a red text</p>
```

Example:

```
<!DOCTYPE html>
<html>
 <head>
  <title>This is page title.</title>
 </head>
 <body>
    <div class="container" style=" padding:
50px; border-radius: 20px; border: 1px solid
black;
    max-width:600px; max-width: auto;">
        <h1> <u> This is an example of Simple
CSS </u> </h1>
        <!-- Here is the Inline Style -->
        <h2 style="color: red"> This is a
heading element </h2>
        <p style="color: purple"> Hello world,
this is a paragraph. </p>
    </div>
 </body>
<html>
```

This is an example of Simple CSS

This is a heading element

Hello world, this is a paragraph.

The inline style sheet.

External style sheets contain CSS instructions, and these are special files, and have a .css file extension. If an external style sheet is installed on any web page, the CSS file will control its sound and appearance.

Using an external style sheet, all your HTML files are linked to a single CSS file to format the pages. This means that if you need to change the design of all your pages, you only need to edit one .css file to make general changes to your entire website. Here are a few reasons why this is better.

- Easier Maintenance

- Reduced File Size

- Reduced Bandwidth

- Improved Flexibility

CSS CLASSES

The class selector lets you style elements in the same HTML element differently. It's similar to what I mentioned in the introduction about inline styles. Except for classes, the style can be overwritten by changing the style sheets. You can use the same class selector over and over in an HTML file.

To put it more simply, the sentence you are reading is defined in my CSS file with the following.

Sample:

```
p
  {
    font-size: small;
    color: #333333
  }
```

It's pretty simple, but let's say we want to change the word "sentence" to green bold text and leave the rest of the sentence untouched. You would do the following to my HTML file.

<p> In simple terms, this <span class="text"> You can write your sentence here </span> in my CSS file you are reading, it is formatted as follows. </p> Then you add the following style selector to my CSS file:

```
.text
  {
  font-size: small;
```

```
  color: #008080;
  font-weight: bold;
}
```

CSS IDs

IDs are similar to classes. Suppose once a specific id has been declared, it cannot be used again within the same HTML file. We commonly use IDs to style the layout elements of a page that will only be needed once, whereas we use classes to style text and such that may be declared various times. The main container for the page is defined by the following example:

```
<div id="container">
```

It is everything within my document is inside this division. </div>. You have chosen the id selector for the "container" division over a class because you only need to use it one time within this file. Then in my CSS file, you will have the following example:

```
#container {
  width: 60%;
  margin: auto;
  padding: 10px;
  border: 2px solid #666;
  background:  #ffffff;
}
```

You will get notice that the id selector should begin with a (#) number sign instead of a (.) period, as the class selector does.

WHAT IS THE DIFFERENCE BETWEEN ID AND CLASS?

We use IDs and Classes to target a particular HTML element to render differently from other similar elements via CSS, but what is the difference?
    Ids are of the "Unique" type:

- Only one ID can be set for each Item.

- The ID should only be used within a page.

You should use the id when you have only one element on the page that you want to render. Your code cannot pass the validation test, which is important for web developers.

Sample:

```
<div id="main-header"> text </div>
```

CSS DIVISION

Previously, you've learned the basics of CSS, how syntax works, and a little bit about classes and ids here. We'll now take a short break from CSS and focus on the HTML side of using it. Sections are block-level HTML elements used to describe sections of an HTML file. A section can contain all the parts that make up your website, including additional sections, spans, images, text, etc. You define a section within an HTML file by placing the following between <body> </body>.

Sample:

```
<body>
<div> Site content go here </div>
</body>
```

You will most likely want to add some style to it. You can do this in the example below:

<div id="container"> Site content go here </div> The CSS file contains:

```
#container
 {
 width: 70%;
 margin: auto;
 padding: 20px;
 border: 1px solid #666;
 background: #ffffff;
}
```

Now everything in this section will be formatted according to the "container" style rule I defined in my CSS file. A split creates a line break by default. You can use both classes and sections of your website.

HOW TO USE CSS PROPERTIES

CSS layouts are used to apply styles to structured documents, such as those created in HTML or XML. CSS layouts are specified at the CSS level. Each property has a set of some possible values. Some properties can affect the type of element, while others apply only to certain groups of elements. The style sheet consists of a list of rules. Each rule or set of rule contains one or more selectors, as well as a declaration block.

CSS features are used within the ad with the corresponding value.

Example:

```
section {
        padding: 20px;
        margin: 20px;
        background-color: cornsilk;
        border: 6px solid gold;
}
h1 {
        color: coral;
}
p {
        color: orange;
}
a:link,
a:visited {
        color: dark orange;
}
a:hover {
        color: orangered;
}
```

Let's discuss various CSS properties in short. We have a separate chapter for each of the properties.

CSS COLORS

Colors are specified using predefined color names, RGB, HEX, HSL, RGBA, HSLA values.

Names

In CSS, a color can be specified by using a pre-defined color name, such as,

1. Tomato

2. Orange

3. Dodger Blue

4. Medium Sea Green

5. Gray

6. SlateBlue

7. Violet

8. LightGray

List of Color Values

Here's a list of color values that can be used with CSS.

- Named Colors

- transparent Keyword

- CS current color Keyword

- 3-Digit Hex Codes

- 4-Digit Hex Codes

- 6-Digit Hex Codes

- CSS 8-Digit Hex Codes

- RGB() Function

- rgba() Function

- HSL() Function

- hsla() Function

- hwb() Function

- System Colors

Here is the complete list of names of colors that are supported by all browsers given below:

1. Alice blue

2. antique white

3. Aqua

4. Aquamarine

5. Azure

6. Beige

7. Bisque

8. Black

9. dark orchid

10. DarkRed

11. DarkSalmon

12. dark turquoise

13. DarkViolet

14. DeepPink

15. DeepSkyBlue

16. Fuchsia

17. Gainsboro

18. GhostWhite

19. Gold

20. HotPink

21. Indian

22. Indigo

23. Ivory

RGB COLOR

There are different colors that add a different atmosphere to various designs. The right choice of colors can make designs and creations look clean, aesthetic, and modern. But the bad colors can make a project difficult for users to interact with because it will look not attractive to others.

The color of the border (border), background (background color), or foreground (color) – the text decorations on the page have a huge impact, so you have to put some effort into getting them right.

CSS lets you use a wide variety of different colors and color systems. They range from named colors to hexagonal colors, RGB() colors, hsl colors, and more.

How to Use RGB Colors in HTML

The easiest way to apply color to your HTML elements is to write your HTML in a .html file. Then, in that file, simply associate your .css style sheet with all the colors and styles you specify.

It makes code easier to read and removes any considered best practice concerns. We can have an about.html file with some HTML code like this:

Example:

```
<!DOCTYPE html>
<html lang="en">
<head>
    <meta charset="UTF-8">
    <meta HTTP-equiv="X-UA-Compatible"
content="IE=edge">
    <meta  content="width=device-width,
name="viewport"initial-scale=1.0">
<style>

h1{
    font-size: 38px;
}


h2 {
/* changes color of text   */
  color: RGB(79, 72, 70);
  text-align:center;
}
.container {
    align-items: center;
    max-width: 800px;
    max-height: auto;
    margin: 280px auto;
    border: 1px solid black;
    border-radius: 20px;
    padding: 50px;
}
```

```
 </style>
 </head>
 <body>
    <div class="container">
       <h1> Use of RGB in CSS </h1>
          <h2> Lorem ipsum sit amet, consectetur
adipiscing elit. Quisque pulvinar lobortis turpis,
ac imperdiet magna. Nulla pretium a sem a luctus.
Pellentesque habitant morbi tristique senectus et
et malesuada fames ac turpis egestas. </h2>
          <h2> Here we use rgb () - rgb(79, 72,
70) </h2>
       </div>
 </body>
 <html>
```

Use of RGB in CSS

Lorem ipsum dolor sit amet, consectetur adipiscing elit. Quisque pulvinar
lobortis turpis, ac imperdiet magna. Nulla pretium a sem a luctus.
Pellentesque habitant morbi tristique senectus et netus et malesuada fames ac
turpis egestas.

Here we use rgb () - rgb(79, 72, 70)

Use of RGB in CSS.

These are the common color user in CSS, you can change its value to create a new color.

Color Value in RGB	Color Name
RGB(0, 255, 255)	aqua
RGB(0, 0, 0)	black
RGB(0, 0, 255)	blue
RGB(255, 0, 255)	fuchsia
RGB(128, 128, 128)	gray
RGB(0, 128, 0)	green
RGB(0, 255, 0)	lime
RGB(128, 0, 0)	maroon
RGB(0, 0, 128)	navy

(Continued)

Color Value in RGB	Color Name
RGB(128, 128, 0)	olive
RGB(128, 0, 128)	purple
RGB(255, 0, 0)	red
RGB(192, 192, 192)	silver
RGB(0, 128, 128)	teal
RGB(255, 255, 255)	white
RGB(255, 255, 0)	yellow

Extended Color Keywords

The following table lists the color keywords defined in the CSS3 specification.

Color Name	RGB Value
aliceblue	RGB(240, 248, 255)
antique white	RGB(250, 235, 215)
aqua	RGB(0, 255, 255)
aquamarine	RGB(127, 255, 212)
azure	RGB(1240, 255, 255)
beige	RGB(245, 245, 220)
bisque	RGB(255, 228, 196)
black	RGB(0, 0, 0)
blanchedalmond	RGB(255, 235, 205)
blue	RGB(0, 0, 255)
blueviolet	RGB(138, 43, 226)
brown	RGB(165, 42, 42)
burlywood	RGB(222, 184, 135)
cadet blue	RGB(95, 158, 160)
chartreuse	RGB(95, 158, 160)
chocolate	RGB(210, 105, 30)
coral	RGB(255, 127, 80)
cornflower blue	RGB(100, 149, 237)
cornsilk	RGB(255, 248, 220)
crimson	RGB(220, 20, 60)
cyan	RGB(0, 255, 255)
darkblue	RGB(0, 0, 139)
dark cyan	RGB(0, 139, 139)
darkgoldenrod	RGB(184, 134, 11)
darkgray	RGB(169, 169, 169)
darkgreen	RGB(0, 100, 0)
dark khaki	RGB(189, 183, 107)
dark magenta	RGB(139, 0, 139)
darkolivegreen	RGB(85, 107, 47)
dark orange	RGB(255, 140, 0)
dark orchid	RGB(153, 50, 204)

(Continued)

Color Name	RGB Value
darkred	RGB(139, 0, 0)
darksalmon	RGB(233, 150, 122)
darkseagreen	RGB(143, 188, 143)
darkslateblue	RGB(72, 61, 139)
darkslategray	RGB(47, 79, 79)
dark turquoise	RGB(0, 206, 209)
dark violet	RGB(148, 0, 211)
deeppink	RGB(255, 20, 147)
deepskyblue	RGB(0, 191, 255)
dim gray	RGB(0, 191, 255)
dodgerblue	RGB(30, 144, 255)
firebrick	RGB(178, 34, 34)
floralwhite	RGB(255, 250, 240)
forest green	RGB(34, 139, 34)
fuchsia	RGB(255, 0, 255)
gainsboro	RGB(220, 220, 220)
ghostwhite	RGB(248, 248, 255)
gold	RGB(255, 215, 0)
goldenrod	RGB(218, 165, 32)
gray	RGB(127, 127, 127)
green	RGB(0, 128, 0)
greenyellow	RGB(173, 255, 47)
honeydew	RGB(240, 255, 240)
hotpink	RGB(255, 105, 180)
indianred	RGB(205, 92, 92)
indigo	RGB(75, 0, 130)
Ivory	RGB(255, 255, 240)
khaki	RGB(240, 230, 140)
lavender	RGB(230, 230, 250)
lavender blush	RGB(255, 240, 245)
lawngreen	RGB(124, 252, 0)
lemon chiffon	RGB(255, 250, 205)
lightblue	RGB(173, 216, 230)
light coral	RGB(240, 128, 128)
light cyan	RGB(224, 255, 255)
lightgoldenrodyellow	RGB(250, 250, 210)
lightgreen	RGB(144, 238, 144)
light grey	RGB(211, 211, 211)
lightpink	RGB(255, 182, 193)
light salmon	RGB(255, 160, 122)
lightseagreen	RGB(32, 178, 170)
lightskyblue	RGB(135, 206, 250)
lightslategray	RGB(119, 136, 153)
lightsteelblue	RGB(176, 196, 222)
light yellow	RGB(255, 255, 224)
lime	RGB(0, 255, 0)
limegreen	RGB(50, 205, 50)

(Continued)

Color Name	RGB Value
linen	RGB(250, 240, 230)
magenta	RGB(255, 0, 255)
maroon	RGB(128, 0, 0)
medium aquamarine	RGB(102, 205, 170)
mediumblue	RGB(0, 0, 205)
medium orchid	RGB(186, 85, 211)
medium purple	RGB(147, 112, 219)
mediumseagreen	RGB(60, 179, 113)
mediumslateblue	RGB(123, 104, 238)
mediumspringgreen	RGB(0, 250, 154)
mediumturquoise	RGB(72, 209, 204)
mediumvioletred	RGB(199, 21, 133)
midnight blue	RGB(25, 25, 112)
mint cream	RGB(245, 255, 250)
misty rose	RGB(255, 228, 225)
moccasin	RGB(255, 228, 181)
navajowhite	RGB(255, 222, 173)
navy	RGB(0, 0, 128)
navy blue	RGB(159, 175, 223)
oldlace	RGB(253, 245, 230)
olive	RGB(128, 128, 0)
olive drab	RGB(107, 142, 35)
orange	RGB(255, 165, 0)
orangered	RGB(255, 69, 0)
orchid	RGB(218, 112, 214)
pale goldenrod	RGB(238, 232, 170)
palegreen	RGB(152, 251, 152)
pale turquoise	RGB(175, 238, 238)
palevioletred	RGB(219, 112, 147)
papayawhip	RGB(255, 239, 213)
peachpuff	RGB(255, 218, 185)
Peru	RGB(205, 133, 63)
pink	RGB(255, 192, 203)
plum	RGB(221, 160, 221)
powderblue	RGB(176, 224, 230)
purple	RGB(128, 0, 128)
red	RGB(255, 0, 0)
rosybrown	RGB(188, 143, 143)
royalblue	RGB(65, 105, 225)
saddle brown	RGB(139, 69, 19)
salmon	RGB(250, 128, 114)
sandy brown	RGB(244, 164, 96)
sea green	RGB(46, 139, 87)
seashell	RGB(255, 245, 238)
sienna	RGB(160, 82, 45)
silver	RGB(192, 192, 192)
skyblue	RGB(135, 206, 235)

(Continued)

Color Name	RGB Value
slate blue	RGB(106, 90, 205)
slate gray	RGB(112, 128, 144)
snow	RGB(255, 250, 250)
spring green	RGB(0, 255, 127)
steelblue	RGB(70, 130, 180)
tan	RGB(210, 180, 140)
teal	RGB(0, 128, 128)
thistle	RGB(216, 191, 216)
tomato	RGB(255, 99, 71)
turquoise	RGB(64, 224, 208)
violet	RGB(238, 130, 238)
wheat	RGB(245, 222, 179)
white	RGB(255, 255, 255)
WhiteSmoke	RGB(245, 245, 245)
yellow	RGB(255, 255, 0)
yellowgreen	RGB(139, 205, 50)

DIFFERENT TYPES OF SELECTORS

There are many different selectors. The examples above use item selectors that select all items of a given type. But we can also make more specific choices. Here are some of the common types of selectors:

Selector Name

- The CSS element selector: This element selector selects HTML elements based on the element name.

  Syntax:

  `<p>`

- The CSS id selector: The selector id uses the id attribute of an element to select a particular element. The id is unique in a single page, so the id is used to select a unique element. To select a particular element with an id, write a hash (#) symbol, followed by the id of the element.

  Syntax:

  `<p id="my-id">` or `<a id="my-id">`

- The CSS class selector: The selector class selects HTML elements with a specific class attribute. To select these elements with a particular class, write a dot (.) character, followed by the class name.

Syntax:

```
<p class="my-class"> Add Some text </p> and
<a class="my-class"> Add Some text</a>
```

- Attribute selector: The elements on the page with the particular attribute.

Syntax:

```
<img src="myimage.png"/>
```

- The CSS universal selector: The universal (*) selects all HTML elements on the page. To select these elements with a specific class, write an asterisk (*) character, followed by the tag name.

Syntax:

```
*
```

Example:

```
{
font-size: 20px:
color: blue;
}
```

- The CSS grouping selector: The grouping selector selects every element with the same style definitions. Now, look at the following code (the h1, h2, and p elements have the same style definitions).

- Pseudo-class selector: The specified elements, but only when in a certain state. (For instance, when a cursor hovers over a link.)

Syntax:

```
a:hover
```

selects <a>, works when the mouse pointer is hovering over the link.

CSS BACKGROUND

The background property allows to control the background of the element. It is a property, which means that it allows to write what would be multiple CSS properties in one.

Example:

```
body {
  background:
      URL(sweettexture.jpg)     /* image */
      top center / 200px 200px /* position / size */
      no-repeat                 /* repeat */
      fixed                     /* attachment */
      padding-box               /* origin */
      content-box               /* clip */
      red;                      /* color */
}
```

There are eight other properties available under the background given below:

- background-repeat
- background-attachment
- background-image
- background-clip
- background-color
- background-position
- background-size
- background-origin

You can use any of the combinations of these properties that you love to use, in almost any order

MULTIPLE BACKGROUNDS

CSS3 added support for various backgrounds, which layer over the top of each other. Any property related to the backgrounds can take a comma (,) separated list, like this:

```
.tagname {
  background: URL(image1.jpg), URL(image2.jpg) black;
  background-repeat: repeat-x, no-repeat;
}
```

CSS BORDER

The border property is a syntax in CSS that accepts multiple values for drawing a line around the element it is applied to:

```
.container{
  border: 3px solid red;
  height: 200px;
  width: 200px;
}
```

Border

The border property accepts more than one of the following values in combination:
border-width: It defines the thickness of the border.

- thin

- medium

- thick

border-style: It specifies the type of the line drawn around the element, including:

- dotted – It defines a dotted border.

- dashed – It defines a dashed border.

- solid – It defines a solid border.

- double – It defines a double border.

- groove – It defines a 3D grooved border. The effect depends on the border-color value.

- ridge – It defines a 3D ridged border. The effect depends on the border-color value.

- inset – It defines a 3D inset border. The effect depends on the border-color value.

- outset – It defines a 3D outset border. The effect depends on the border-color value.

- none – It defines no border.

- hidden – It defines a hidden border.

border-color: It specifies the color of the border and accepts all valid color values.

CSS BORDER WIDTH AND COLOR

The border-width property defines the width of the four borders. The width can also be set as a specific size (such as in px, pt, cm, em, etc.) or by using the three pre-defined values such as thin, medium, or thick.

Example:

```
p.one {
  border-style: solid;  //You can add your border
width here
  border-width: 5px;
}
```

The border color property is used to define the color of the four borders. The color can be set by:

- name – It specifies a color name, like "red."

- HEX – It specifies a HEX value, like "#ff0000."

- RGB – It specifies an RGB value, like "rgb(255,0,0)."

- HSL – It specifies an HSL value, like "hsl(0, 90%, 50%)."

- Transparent

CSS MARGINS

CSS genetic features are used to create space around elements, without any defined parameters. With CSS, you have full control over margins. There are features for setting the margin for each element of the element such as top, right, bottom, and left. CSS has features that specify a margin for each element:

- margin-top

- margin-right

- margin-bottom

- margin-left

All genetic features can have the following values:

- default – The browser calculates limit.

- length – It specifies margin in px, pt, cm, etc.

- % – It specifies margin in% of the width of the content.

- inheritance – it specifies that the margin should inherit from the parent.

CSS PADDING

CSS finishing features are used to generate space near element content, within any defined parameters. With CSS, you have complete control over the pads. There are pads setup properties on each side of the element (top, right, bottom, and left). CSS has paddling specifications for each element of the element:

- padding-top

- padding-right

- padding-floor

- padding-left

All completion buildings may have the following values:

- length – It specifies padding in px, pt, cm, etc.

- % – It specifies padding in% of the width of the content.

- inheritance – It specifies that padding should inherit from the parent element.

CSS BOX MODEL

This model is a box that wraps everything HTML contains: margins, borders, padding, and original content. In CSS, the term "box model" is used when referring to design and structure.

Description of different parts:

- Content – The contents of the box, in which the text and images appear.

- Padding – It clears the area around the content. The padding is obvious.

- Border – A border around padding and content.

- Margin – It can clear an area outside of the border. The margin is clear.

CSS TABLES

The table in CSS is used to apply various style elements to HTML Table elements to organize data in rows and columns or in a more complex format. Tables are used in communication and research data analysis. The table layout in CSS can be used to display table layouts. This feature is used primarily to set the algorithm used to edit <table> cells, rows, and columns.

Border: Used to specify parameters in a table.

Syntax:

```
border: table_width table_color;
```

CSS FONTS

The four sides of an element can be set side by side such as margin-top, margin-right, margin-bottom, margin-left, padding-top, padding-right, padding-bottom, and padding-left are the self-explanatory properties you can use. You can set the following font properties of an element:

- The font family property is used to change the writing of a font.

- The font style property is used to make a font italic or oblique.

- The font variant property is used to create a small-caps effect.

- The font weight property is used to increase or decrease how bold light a font appears.

- The font size property is used to increase or decrease the size of a font.

- The font property is used to specify the number of other font properties.

CSS TEXT

- The property color is used to set the color of a text.

- The property direction is used to set the text direction.

- The letter-spacing property is used to add or subtract space between letters that make a word.

- The word-spacing property is used to add or subtract space between words of a sentence.

- The text indent property is used to indent the content of a paragraph.

- The text align property is used to align the content of a document.

- The text decoration property is used to underline, overline, and strikethrough text.

- The text-transform property is used to capitalize content or content text to uppercase or lowercase letters.

- The white space property is used to control the flow and formatting of text.

- The text-shadow property is used to set text shadow around a text.

CSS FUNCTIONS

CSS functions are used as a number of different CSS structures. For example, you can use RGB () function to assign a color value (similar to the color: RGB (205, 0, 215)), the attr () function to retrieve the HTML attribute value. Many functions are used in CSS conversion. For example, a rotating function () could be used to rotate an element, a scale () function could be used to resize an object, and a translate() function can be used to move an element.

There are other useful functions, such as a circle () to attach an object to a circle or to create a circle for text to flow around, as well as a counting function () that can be used to give a calculated value to a plot.

Here is a list of CSS functions in CSS3.

- attr()

- blur()

- brightness()
- calc()
- circle()
- contrast()
- counter()
- counters()
- cubic-bezier()
- drop-shadow()
- ellipse()
- filter()
- grayscale()
- HSL()
- hsla()
- hue-rotate()
- hwb()
- image()
- inset()
- invert()
- linear-gradient()
- matrix()
- matrix3d()
- opacity()
- perspective()
- polygon()
- radial-gradient()
- repeating-linear-gradient()

- repeating-radial-gradient()
- RGB()
- rgba()
- rotate()
- rotate3d()
- rotateX()
- rotateY()
- rotateZ()
- saturate()
- sepia()
- scale()
- scale3d()
- scaleX()
- scaleY()
- scaleZ()
- skew()
- skewX()
- skew()
- symbols()
- translate()
- translate3d()
- translateX()
- translateY()
- translateZ()
- URL()
- var()

WHY FLEXBOX?

For a long time, the only reliable browser-compatible tools available for creating layouts were features such as floats and positioning. These work, but in some ways they are limited and frustrating.

The following building designs are difficult or impossible to achieve with such tools in any form of simple, flexible approach:

- Vertically centering a block element of content inside its parent.

- Making all the children of a container take up the same amount of the width/height.

- Making all columns in a multiple-column layout with the same height even if it contains a different amount of content.

CSS FLEXBOX

The flexbox CSS layout allows you to easily format HTML. Flexbox makes it easy to direct objects horizontally and horizontally using lines and columns. Items will "change" into different sizes to fill the space. It makes responsive design easier.

CSS flexbox is ideal for standard website design or your application. It is easy to read, supports all modern browsers, and does not take much time to get the basics right. By the end of this same section, you will be ready to start using flexbox for your web projects.

Flexbox is a 1D layout for arranging objects in rows or columns. Items are flexible to fill in the blanks or to shrink into smaller spaces.

The flex container properties are:

- flex-direction

- flex-wrap

- flex-flow

- justify-content

- align-items

- align-content

CSS MEDIA TYPES

INTRODUCTION TO MEDIA TYPES

One of the most important features of stylesheets is that you can specify different style sheets for different types of media. This is a good way to create Web pages ready for printer – It just provides a separate style sheet for the "print" media type.

Some CSS features are for media only. For example, the page sorting feature only applies to pages with pages. There are other few things that can be shared by different media types, but they may require different values for that feature. The font size feature, for example, can be used for screen and print media, but probably at different values.

Documentation often requires a large font on a computer screen compared to paper for better reading, and sans-serif fonts are considered easy to read on screen, while serif fonts are famous for printing. It is therefore necessary to clarify that a style sheet, or set of style rules, applies to certain types of media.

METHOD 1: USING @MEDIA AT-RULES

The @media rule is used to define different style rules for different media in a single style sheet. It is followed by a comma separated list of media types and a CSS declaration block containing targeted media style rules.

Example: The style tells the browser to display body content in 14 pixels Arial font on the screen, but when printing it will be in Times font 12 points.

```
@media screen {
    body {
        color: black;
        font-size: 12px;
    }
}
@media print {
    body {
        color: # fff;
        font-size: 10pt;
    }
}
```

```
@media screen, print {
    body {
        line-length: 1;
    }
}
```

METHOD 2: USING @IMPORT AT-RULES

The @import rule is another way to specify specific media style details – It just specifies comma separated media types after the URL of the imported style sheets.

Example:

```
@import URL ("CSS / screen.css") screen;
@import URL ("CSS / print.css") print;
div {
    background: # f5f5f5;
    line length: 1.2;
}
```

METHOD 3: USING THE <link> ELEMENT

The media attribute in the <link> feature is used to specify targeted media on an external style sheet within an HTML document.

```
<link rel = "stylesheet"  href = "css / common.css"
media = "all">
<link rel = "stylesheet"  href = "css / print.css"
media = "print">
```

DIFFERENT MEDIA TYPES

The following lists the various media types that are used to target various types of devices such as printers, devices, computer screens, etc.

- all: It is used for all media-type devices.

- aural: It is used for speech and sound synthesizers.

- braille : It is used for tactile feedback devices.

- embossed: It is used for paged braille printers.

- handheld: It is used for small or handheld devices small screen devices such as mobile phones or PDAs.

- print: It is used for printers.

- projection: It is used for projected presentations, for example, projectors.

- screen: It is used primarily for color computer screens.

- tty: It is used for media using a fixed pitch character grid such as teletypes, terminals, and portable devices with limited display capabilities.

- tv: It is used for television-type devices with low resolution, color, limited scrollability screens, and sound available.

MEDIA QUESTIONS AND RESPONSIVE WEB DESIGN

Media queries allow to customize the presentation of web pages to a specific range of devices such as mobile phones, tablets, desktops, etc., without change in markups. A media query contains media type and zero or more expressions such as the type and conditions of certain media features such as device width or screen resolution. Here's a simple example of a typical media query for standard devices.

MEDIA QUERIES

Responsive design was able to emerge due to media query. The definition of Media Queries Level 3 was Candidate Recommendations in 2009, which means it was considered ready for implementation in browsers. Media queries allow us to create a series of tests and apply CSS by choosing to customize the page style to the user's needs.

For example, the following media quiz questions to see if the current web page is displayed as screen media (so it is not a printed document) and the viewing area is 800 pixels wide. CSS selector .container will only be used if these two are true.

```
/* Smartphones (portrait and landscape) ---------- */
@media screen and (min-width: 320px) and (max-width:
480px) {
    /* styles */
}
/* Smartphones (portrait) ---------- */
@media screen and (max-width: 320px) {
    /* styles */
}
```

```
/* Smartphones (landscape) ---------- */
@media screen and (min-width: 321px){
    /* styles */
}
/* Tablets, iPads (portrait and landscape)
---------- */
@media screen and (min-width: 768px) and (max-width:
1024px){
    /* styles */
}
/* Tablets, iPads (portrait) ---------- */
@media screen and (min-width: 768px){
    /* styles */
}
/* Tablets, iPads (landscape) ---------- */
@media screen and (min-width: 1024px){
    /* styles */
}
/* Desktops and laptops ---------- */
@media screen and (min-width: 1224px){
    /* styles */
}
/* Large screens ---------- */
@media screen and (min-width: 1824px){
    /* styles */
}
```

BENEFITS YOU MAY KNOW

Let's learn more about CSS benefits such as:

1. One of the main benefits of CSS is that CSS is very useful for Digital Marketing purposes where every promotion of your web page provides a business opportunity. Therefore, more content with a code formula with the help and use of CSS techniques is better for digital marketing and your business in the end.

2. CSS offers its users, by making small changes to your website, similar changes are seen in other parts of the website. The bigger your website, the more content and layouts are where CSS saves you a lot of time when it comes to conversions. It also checks and verifies that each page is consistent with the CSS benefits.

3. W3C works with the goal of making the entire website equally accessible to disabled users the way normal users use it. First CSS gains access by creating layout and document making it easy for any screen reader to read content accurately. This is useful for deaf users as they rely on the screen reader to run web applications.

4. For a website to work properly, it must have a fast download time. Nowadays, people usually wait a few seconds for a website to load. Therefore, it is important to ensure a fast pace. For companies looking to ensure fast and smooth website information, CSS becomes key to their success.

5. CSS is easy to maintain due to the short maintenance time. This is because single-line code conversion affects the entire web page. Also, if upgrades are needed, a little more effort is needed to minimize changes to the webpage code.

6. CSS empowers developers to ensure that style features are applied consistently across every few web pages.

7. CSS saves a lot of time and effort in the web development process due to fast loading time. Here, a little time ensures the good performance of the designer.

CSS CONS

There are a few drops while using CSS. One has to be aware of these bad situations in order to know and take care of them while designing a website.

1. Confusion due to too many CSS levels: Beginners are at high risk in this regard. They may be confused when choosing to read CSS as there are many CSS levels like CSS2, CSS3, etc.

2. Browser problems: Different browsers work differently. Therefore, you should check if the changes made to the website by the CSS codes are visible to all browsers.

3. Security issues: Security is critical in today's world driven by technology and data. One of the bad things about CSS is that it has limited security.

4. Additional developer work: Design services are required to consider and test all CSS codes in all different browsers for compatibility. As a result of testing developer compatibility in different browsers, their workload is growing.

All in all, we can say that if you like web development, try learning HTML and CSS. For device compatibility, read the Bootstrap framework. While you may see some disadvantages of CSS, many of the benefits are counter-productive and make sure your web development process is smooth and efficient.

CHAPTER SUMMARY

Here we have covered all the basic concepts of CSS that everyone can use to make their pages attractive. So important thing for working with CSS is just only practice, once you start practicing it, you will get to the point easily. Then in the coming chapter, you will learn CSS properties that make your work so easy.

CSS Properties

IN THIS CHAPTER

➢ Introduction

➢ CSS Layouts (Properties)

➢ CSS Rules

➢ The @keyframes Rule

➢ CSS Background Properties

➢ CSS Border Properties

In the last chapter, we learned about the basics of CSS, its history, types, versions, and other properties such as how to add CSS on classes, id, commenting out code, basic HTML, and linking HTML with CSS. Now here we are going to discuss various important properties of CSS such as display, positioning, border, grid, flex height, width, color, font, and so on.

You style HTML elements via CSS properties. Different HTML elements may have different CSS layouts to set. CSS layouts can be organized into CSS rules. CSS rule binds a set of CSS layouts together, and applies all the elements to HTML elements that match the CSS rule. This document will include both CSS layouts and CSS rules in more detail.

In CSS, there are various properties; below you will get the list of all attributes and then we separate those properties into different sections.

DOI: 10.1201/9781003358060-2

1. align-content
2. align-items
3. align-self
4. all
5. animation
6. animation-delay
7. animation-direction
8. animation-duration
9. animation-fill-mode
10. animation-iteration-count
11. animation-name
12. animation-play-state
13. animation-timing-function
14. azimuth
15. backface-visibility
16. background-attachment
17. background-blend-mode
18. background-clip
19. background-color
20. background-image
21. background-origin
22. background-position
23. background-repeat
24. background-size
25. background
26. bleed
27. border-bottom-color
28. border-bottom-left-radius
29. border-bottom-right-radius
30. border-bottom-style
31. border-bottom-width
32. border-bottom
33. border-collapse
34. border-color
35. border-image
36. border-image-outset
37. border-image-repeat
38. border-image-source
39. border-image-slice
40. border-image-width
41. border-left-color
42. border-left-style
43. border-left-width
44. border-left
45. border-radius
46. border-right-color
47. border-right-style
48. border-right-width
49. border-right
50. border-spacing
51. border-style
52. border-top-color
53. border-top-left-radius
54. border-top-right-radius
55. border-top-style
56. border-top-width
57. border-top
58. border-width
59. border
60. bottom
61. box-decoration-break
62. box-shadow
63. box-sizing
64. break-after
65. break-before
66. break-inside
67. caption-side
68. caret-color
69. clear
70. clip
71. color
72. color-interpolation-filters
73. columns
74. column-count
75. column-fill
76. column-gap
77. column-rule
78. column-rule-color
79. column-rule-style
80. column-rule-width
81. column-span
82. column-width
83. content
84. counter-increment
85. counter-reset
86. cue-after
87. cue-before
88. cue
89. cursor
90. direction
91. display
92. elevation
93. empty-cells
94. filter
95. flex
96. flex-basis

(Continued)

97. flex-direction
98. flex-flow
99. flex-grow
100. flex-shrink
101. flex-wrap
102. float
103. flood-color
104. flood-opacity
105. font-family
106. font-feature-settings
107. font-kerning
108. font-language-override
109. font-size-adjust
110. font-size
111. font-stretch
112. font-style
113. font-synthesis
114. font-variant
115. font-variant-alternates
116. font-variant-caps
117. font-variant-east-asian
118. font-variant-ligatures
119. font-variant-numeric
120. font-variant-position
121. font-weight
122. font
123. gap
124. grid-area
125. grid-auto-columns
126. grid-auto-flow
127. grid-auto-rows
128. grid-column-end
129. grid-column-gap
130. grid-column-start
131. grid-column
132. grid-gap
133. grid-row-end
134. grid-row-gap
135. grid-row-start
136. grid-row
137. grid-template-areas
138. grid-template-columns
139. grid-template-rows
140. grid-template
141. grid
142. hanging-punctuation
143. height
144. hyphens
145. image-rendering
146. isolation
147. justify-content
148. justify-items
149. justify-self
150. left
151. letter-spacing
152. lighting-color
153. line-break
154. line-height
155. list-style-image
156. list-style-position
157. list-style-type
158. list-style
159. margin-bottom
160. margin-left
161. margin-right
162. margin-top
163. margin
164. marker-offset
165. marks
166. max-height
167. max-width
168. min-height
169. min-width
170. mix-blend-mode
171. nav-up
172. nav-down
173. nav-left
174. nav-right
175. object-fit
176. object-position
177. opacity
178. order
179. orphans
180. outline-color
181. outline-offset
182. outline-style
183. outline-width
184. outline
185. overflow
186. overflow-wrap
187. overflow-x
188. overflow-y
189. padding-bottom
190. padding-left
191. padding-right
192. padding-top
193. padding
194. page-break-after

(Continued)

195. page-break-before
196. page-break-inside
197. page
198. pause-after
199. pause-before
200. pause
201. perspective
202. perspective-origin
203. pitch-range
204. pitch
205. place-content
206. place-items
207. place-self
208. play-during
209. position
210. quotes
211. resize
212. rest-after
213. rest-before
214. rest
215. richness
216. right
217. row-gap
218. size
219. speak-header
220. speak-numeral
221. speak-punctuation
222. speak
223. speech-rate
224. stress
225. tab-size
226. table-layout
227. text-align
228. text-align-all
229. text-align-last
230. text-combine-upright
231. text-decoration
232. text-decoration-color
233. text-decoration-line
234. text-decoration-skip
235. text-decoration-style
236. text-justify
237. text-indent
238. text-orientation
239. text-overflow
240. text-shadow
241. text-transform
242. text-underline-position
243. top
244. transform
245. transform-box
246. transform-origin
247. transform-style
248. transition
249. transition-delay
250. transition-duration
251. transition-property
252. transition-timing-function
253. unicode-bidi
254. vertical-align
255. visibility
256. voice-balance
257. voice-duration
258. voice-family
259. voice-pitch
260. voice-range
261. voice-rate
262. voice-stress
263. voice-volume
264. volume
265. white-space
266. widows
267. width
268. will-change
269. word-break
270. word-spacing
271. word-wrap
272. writing-mode
273. z-index

CSS LAYOUTS (PROPERTIES)

The CSS feature is used to set the style of HTML objects. The CSS layout consists of two parts, the name of the property and the value of the property. Property value included between two (""quotes).

CSS style styles are part of the HTML object. Here are a few examples:

```
<div style = "border: 1px solid black; font size:
18px; "> Style is an attribute of division tag </div>
```

In this example, two CSS features are used in the div feature: border and font size structures.

The CSS layout declaration contains the layout name and layout value. The name of the structure comes first, then the colon, then the value. Here a common pattern for the following CSS property-value is given below:

```
Property-name: property-value
```

If you specify more than one CSS property, each word – the cooked value is divided into a semicolon as follows:

```
property1: property-value;
property2: property-value;
```

The final local declaration does not have to end with a semicolon, but it makes it easy to add more CSS features without forgetting to add that extra semicolon. There are many CSS features that you can specify in different HTML elements. These CSS layouts are covered in their text.

CSS RULES

CSS rule is a set of one or more CSS elements to be applied to one or more targeted HTML elements. CSS rule contains a CSS selector and a set of CSS layouts. The CSS selector determines which HTML elements should be targeted by the CSS rule. CSS features specify what styles of HTML objects are intended.

This example creates a CSS rule that governs all div features, as well as a set of CSS locations and font size for target objects. The part of the CSS selector CSS rule is the front part {. In the above example, it is part of the div of the CSS rule. CSS layouts are listed within the block {...}. CSS rules must be specified within the style element or within the external CSS file.

Here is an example of CSS rule:

```
div {
    border: 1px solid black;
      font size: 18px;
}
```

The following section consists of the complete list of standard CSS properties belonging to the latest CSS3 specifications. All the properties are grouped into categories given below:

- CSS animation properties: The animation property is a CSS property for animation-name, animation-duration, animation-direction, animation-timing-function, animation-delay, animation-iteration-count, animation-fill-mode, and animation-play-state. Here is the default value of the animation property:

```
none 0 ease 0 1 with normal none running
```

It applies to all elements, before and ::after (other pseudo-elements). The following table describes the values of animation property.

1. animation-name property: The animation-name CSS property defines the name of @keyframes defined animations that should be applied to the selected element. The syntax of the property is given below:

```
animation-name:  keyframe name | none | initial |
inherit
```

2. animation-duration property: The animation-duration CSS property defines the number of seconds or milliseconds animation must take to complete one cycle. The syntax of this property is given as:

```
animation-duration: time | initial | inherit
```

3. animation-timing-function property: : The animation-timing-function CSS property specifies how a CSS animation should progress over the duration of each cycle. The syntax of this property is given as:

```
animation-timing-function: linear | ease | ease-in
| ease-out | ease-in-out | cubic-bezier(n,n,n,n) |
initial | inherit
```

4. animation-delay: It specifies a delay before the animation will start. The animation-delay CSS property defines when the animation will start. The value of this property can be specified in seconds (s) or milliseconds (ms).

The syntax of this property is given as:

```
animation-delay:   time | initial | inherit
```

5. animation-iteration-count property: The animation iteration count property defines the number of times an animation cycle should play before stopping.

The syntax of this property is given as:

```
animation-iteration-count:  number | infinite |
initial | inherit
```

6. animation-direction property: The animation-direction CSS property specifies whether the animation should be played in reverse on alternate cycles or not.

The syntax of this property is given as:

```
animation-direction:  normal | reverse | alternate
| alternate-reverse | initial | inherit
```

7. animation-fill-mode property: The animation-fill-mode property defines how a CSS animation should apply styles to the target before and after it is executing.

The syntax of this property is given as:

```
animation-fill-mode: none | backwards | forwards
| both | initial | inherit
```

The following table defines the values of this property:

Value	Description
none	The animation will not apply to the target before or after it is executing.
forwards	After the animation ends, the animation will apply the values for the time the animation ended.
backwards	The animation will apply the values defined in the keyframe that will start the 1st iteration of the animation, during the period defined by the property. Either the values of the from keyframe or to those of the to keyframe.

8. animation-play-state property: The animation-play-state property specifies whether an animation is playing or paused.

The syntax of this property is given as:

```
animation-play-state:  paused | running | initial
| inherit
```

9. initial: It sets the property to its default value.

10. inherit: If it specifies, the associated element takes the calculated value of its parent element's animation property.

THE @KEYFRAMES RULE

When we specify CSS styles inside the @keyframes rule, the animation will change from the current style to new style at certain times.

Example:

```
<html>
  <head>
    <title> Title </title>
    <style>
    * {
  margin: 0;
  padding: 0;
}

.banner-text {
  width: 100%;
  position: absolute;
  z-index: 1;
}

.banner-text h2 {
  text-align: center;
  color: #fff;
  font-size: 50px;
  margin-top: 20%;
}
.animation-area {
  background: #fdc830;
  background: -webkit-linear-gradient(to right,
#f37335, #fdc830);
  background: linear-gradient(to right, #f37335,
#fdc830);
  width: 100%;
  height: 100vh;
}
.box-area {
```

```css
  position: absolute;
  top: 0;
  left: 0;
  width: 100%;
  height: 100%;
  overflow: hidden;
}
.box-area li {
  position: absolute;
  display: block;
  list-style: none;
  width: 25px;
  height: 25px;
  background: rgba(255, 255, 255, 0.2);
  animation: animate 20s linear infinite;
  bottom: -150px;
}
.box-area li:nth-child(1) {
  left: 86%;
  width: 80px;
  height: 80px;
  animation-delay: 0s;
}
.box-area li:nth-child(2) {
  left: 12%;
  width: 30px;
  height: 30px;
  animation-delay: 1.5s;
  animation-duration: 10s;
}
.box-area li:nth-child(3) {
  left: 70%;
  width: 100px;
  height: 100px;
  animation-delay: 5.5s;
}
.box-area li:nth-child(4) {
  left: 42%;
  width: 150px;
  height: 150px;
  animation-delay: 0s;
  animation-duration: 15s;
}
```

```css
.box-area li:nth-child(5) {
  left: 65%;
  width: 40px;
  height: 40px;
  animation-delay: 0s;
}
.box-area li:nth-child(6) {
  left: 15%;
  width: 110px;
  height: 110px;
  animation-delay: 3.5s;
}
@keyframes animate {
  0% {
    transform: translateY(0) rotate(0deg);
    opacity: 1;
  }
  100% {
    transform: translateY(-800px) rotate(360deg);
    opacity: 0;
  }
}

  </style>
  </head>
  <body>
    <div class="banner-text">
      <h2>Animated background</h2>
    </div>
    <div class="animation-area">
      <ul class="box-area">
        <li></li>
        <li></li>
        <li></li>
        <li></li>
        <li></li>
        <li></li>
      </ul>
    </div>
  </body>
</html>
```

The output of the code is given below:

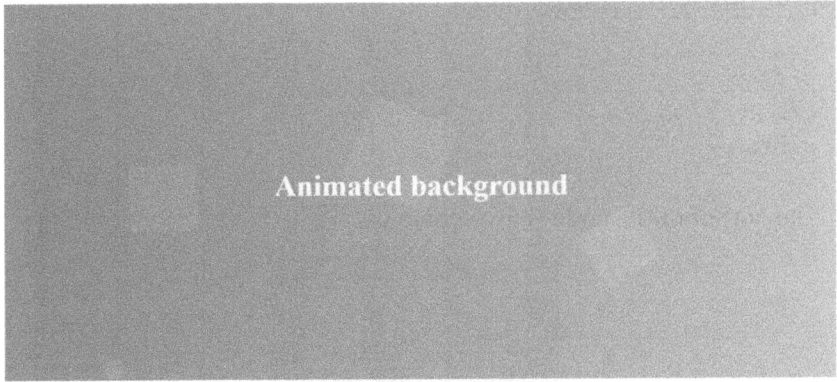

CSS animation property.

Another example:

```
<!DOCTYPE html>
<html>
<head>
<style>
.container {
  display:Flex;
  width: 200px;
  height: 200px;
  background-color: red;
  animation-name: example;
  animation-duration: 4s;
  text-align: center;
  align-items: center;
  margin:0 auto;
}

@keyframes example {
    0%   {background-color: rgb(248, 169, 169);}
    25%  {background-color: rgb(197, 197, 92);}
    50%  {background-color: rgb(118, 118, 193);}
    100% {background-color: rgb(135, 245, 135);}
}
</style>
</head>
```

```
<body>
<div class="container">
<h1>CSS Animation</h1>

</div>
</body>
</html>
```

The output of the code is given below:

CSS animation property (second example).

Now in the below example, we are going to use animation with its properties that we discussed above.

```
<!DOCTYPE html>
<html>
<head>
<style>

.demo_container{
  text-align: center;
}
.parent {
display: flex;
align-items: center;
justify-content: center;
}
.child {
  width: 100px;
  height: 100px;
  background-color: rgb(111, 207, 236);
  position: relative;
  animation-name: demo;
```

```
   animation-duration: 4s;
   animation-delay: 2s;
   display: flex;
align-items: center;
justify-content: center;
border-radius:20px
}

@keyframes demo {
   0%   { background-color: rgb(234, 141, 141); left:
-100px; top: 0px; }
   25%  { background-color: rgb(212, 212, 156); left:
200px; top: 0px; }
   50%  { background-color: rgb(103, 103, 186); left:
200px; top: 200px; }
   75%  { background-color: rgb(141, 230, 141); left:
-100px; top: 200px;}
   100% { background-color: rgb(199, 99, 99); left:
-100px; top: 0px; }
}
</style>
</head>
<body>
<div class="demo_container">
  <h1>CSS Animation</h1>

  <div class="parent">
    <div class="child"></div>
    </div>

</div>
</body>
</html>
```

The output of the code is given below:

CSS Animation

CSS animation property (third example).

CSS BACKGROUND PROPERTIES

The background is a shorthand property for setting a single background property, that is, background-image, background-position, background-size, background-repeat, background-attachment, background-origin, background-clip, and background-color in a single declaration.

The syntax of this property is given as:

```
background:   [ the image position/size repeat
attachment the origin clip color ] | initial | inherit
```

1. background-attachment property: The background-attachment CSS property defines whether the background image scrolls with the document or remains fixed to the viewing area. The syntax of this property is given as:

   ```
   background-attachment:   scroll | fixed | initial |
   inherit.
   ```

2. background-color property: The background-color property sets the background color of an element. You can set the color of the background either through a color value or the keyword transparent.

   The background of any element is the whole size of the element, including padding and border (not the margin). See the box model. The syntax of this property is given as:

   ```
   background-color:   color | transparent | initial
   | inherit
   ```

3. background-clip property: The background-clip property specifies whether an element's background, either the color or image, extends it's border or not. The syntax of this property is given as:

   ```
   background-clip:   border-box | padding-box |
   content-box | initial | inherit
   ```

Example:

```
<!DOCTYPE html>
<html lang="en">
<head>
```

```
<meta charset="utf-8">
<title>Example of CSS3 Background Clipping</
title>
<style>
  .box {
    width: 200px;
    height: 100px;
    padding: 10px;
    border: 2px dashed #333;
    background: rgb(140, 190, 202);

  }
  .clip1 {
    background-clip: border-box;
  }
  .clip2 {
    background-clip: padding-box;
  }
  .clip3 {
    background-clip: content-box;
  }
</style>
</head>
<body>

  <h1> CSS Background Clip </h1>
    <div class="flex-1">
      <h2>Default Background Behavior </h2>
      <div class="box"></div>
      <h2>Background Clipping Using border-box
</h2>
      <div class="box clip1"></div>
    </div>
    <div class="flex-2">
    <h2>Background Clipping Using padding-box
</h2>

    <div class="box clip2"></div>
    <h2>Background Clipping Using content-box
</h2>
    <div class="box clip3"></div>
    </div>
</body>
</html>
```

The output of the code is given below:

CSS Background Clip

Default Background Behavior

Background Clipping Using border-box

Background Clipping Using padding-box

Background Clipping Using content-box

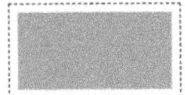

CSS background clip.

4. background-image property: The background-image property sets the background image for an element. It is often more good to use the shorthand background property.

The syntax of the property is given as:

```
background-image:   url | none | initial | inherit
```

Example:

```
<!DOCTYPE html>
<html>
<head>
<style>

.demo_container{
  text-align: center;
}
```

```
.parent {
display: flex;
align-items: center;
justify-content: center;
background-image: url("/images-1.jpg");

}
.child {
  height:100vh;
  display: flex;
align-items: center;
justify-content: center;
}

</style>
</head>
<body>
<div class="demo_container">

  <div class="parent">
    <div class="child"> <h1>CSS  Background
Image </h1></div>
    </div>

</div>
</body>
</html>
```

The output of the code is given below:

CSS background image.

5. background-origin property: The background-origin CSS property specifies the positioning area of the background, that is, the position of the origin of an image specified using the background-image property. The syntax of this property is given as:

```
background-origin:  border-box | padding-box |
content-box | initial | inherit
<!DOCTYPE html>
<html>
<head>
<style>

.demo_container{
  text-align: center;
}
.parent {
  height:500px;
display: flex;
align-items: center;
justify-content: center;

}
.child {
  width: 500px;
    height:400px;
        padding: 10px;
        border: 6px dashed #333;
background: url("/images-1.jpg") no-repeat;
    background-size: cover;
        background-origin: content-box;
}

</style>
</head>
<body>
<div class="demo_container">

  <div class="parent">
    <div class="child"> <h1>CSS  Background Origin
- Here in this you can apply padding  </h1></div>
    </div>

</div>
</body>
</html>
```

The output of the code is given below:

CSS background origin.

Padding box example:

```
<!DOCTYPE html>
<html>
<head>
<style>

.demo_container{
  text-align: center;
}
.parent {
  height:500px;
display: flex;
align-items: center;
justify-content: center;

}
.child {
  width: 500px;
    height:400px;
        border: 6px dashed #333;
background: url("/images-1.jpg") no-repeat;
    background-size: cover;
        background-origin: padding-box  ;
}
```

```
</style>
</head>
<body>
<div class="demo_container">

  <div class="parent">
    <div class="child"> <h1>CSS  Background Origin
- Here in this you can apply padding  </h1></div>
    </div>

</div>
</body>
</html>
```

The output of the code is given below:

CSS background origin – padding box.

6. background-position property: The background-position property sets the initial position (i.e., origin) of the element's background image. It is often more good to use the shorthand background property.

Its default value is : 0% 0%.

The syntax of this property is given as:

```
background-position: [ percentage | length | left
| center | right ]1 or 2 values | initial |
inherit
```

Example:

```
<!DOCTYPE html>
<html>
<head>
<style>

.demo_container{
  text-align: center;
}
.parent {
display: flex;
align-items: center;
justify-content: center;

}
.child {
        border: 6px dashed #333;
background: url("/images-1.jpg") no-repeat;
    background-position: 50% center;

}

</style>
</head>
<body>
<div class="demo_container">

  <div class="parent">
    <div class="child"> <h1>CSS  Background
Position - 50% center </h1></div>
    </div>

</div>
</body>
</html>
```

The output of the code is given below:

CSS background position.

7. background-repeat property: The background-repeat CSS property specifies whether the background image is repeated or tiled after it has been sized and positioned. It is often good to use the shorthand background property.

The syntax of the property is given as:

```
background-repeat: repeat | repeat-x | repeat-y |
no-repeat | initial | inherit
```

Example:

```
<!DOCTYPE html>
<html>
<head>
<style>

.demo_container{
   text-align: center;
}
.parent {
display: flex;
align-items: center;
justify-content: center;

}
.child {
   width:500px;
   height:500px;
         border: 6px dashed #333;
background: url("/images-1.jpg") no-repeat;
     background-repeat: repeat-y;

}

</style>
</head>
<body>
<div class="demo_container">
   <div class="parent">
     <div class="child"> <h1> CSS Background
Repeat - repeat-y </h1></div>
     </div>

</div>
</body>
</html>
```

The output of the code is given below:

CSS background repeat(y).

Another example:

```
<!DOCTYPE html>
<html>
<head>
<style>

.demo_container{
  text-align: center;
}
.parent {
display: flex;
align-items: center;
justify-content: center;

}
.child {
  width:500px;
  height:500px;
       border: 6px dashed #333;
background: url("/images-1.jpg") no-repeat;
    background-repeat: repeat-x;

}
```

```
</style>
</head>
<body>
<div class="demo_container">

  <div class="parent">
    <div class="child"> <h1>CSS  Background
Repeat - repeat-x </h1></div>
    </div>

</div>
</body>
</html>
```

The output of the code is given below:

CSS background repeat(x).

8. background-size property: The background-size property defines the size of the background images.

The syntax of the property is given as:

```
background-size: length attribute| percentage
attribute| auto  attribute | cover attributes
attribute | contain attribute| initial attribute|
inherit attribute
```

Example:

```
<!DOCTYPE html>
<html>
```

```
<head>
<style>

.demo_container{
  text-align: center;
}
.parent {
display: flex;
align-items: center;
justify-content: center;

}
.child-1 {
  width:500px;
  height:500px;
        border: 6px dashed #333;
background: url("/images-1.jpg") no-repeat;
background-size: contain;

}
.child-2 {
  width:500px;
  height:500px;
        border: 6px dashed #333;
background: url("/images-1.jpg") no-repeat;
background-size: cover;

}


</style>
</head>
<body>
<div class="demo_container">

  <div class="parent">
    <div class="child-1"> <h1>CSS  Background
Size - Contain </h1></div>
    <div class="child-2"> <h1>CSS  Background
Size - Cover </h1></div>

    </div>
```

```
</div>
</body>
</html>
```

The output of the code is given below:

CSS background size – contained cover and full cover.

CSS BORDER PROPERTIES

The border property sets the width, style, and color for all four sides of an element's border. It is a shorthand property for setting the individual border properties, that is, border-width, border-style, and border-color in a single declaration.

The syntax of the property is given as:

```
border:   [ border-width border-style border-color ] |
initial | inherit
```

Example:

```
<!DOCTYPE html>
<html>
<head>
<style>

.demo_container{
  text-align: center;
  display: flex;
  justify-content: center;
  align-items: center;

}
```

```
.child-1 {
  width:500px;
  height:500px;
        border: 5px solid rgb(209, 21, 21);
        background-color: rgb(226, 230, 12);
        color:white;
        justify-content: center;
        align-items: center;
        font-size: 32px;

}


</style>
</head>
<body>
<div class="demo_container">
    <div class="child-1"> <h1>CSS  Border </h1>
</div>
</div>
</body>
</html>
```

The output of the code is given below:

CSS border property.

1. border-width property: The border-width CSS property is a short-hand property for setting a singleborder width property, that is, border-top-width, border-right-width, border-bottom-width, and border-left-width in a single declaration. The syntax of this property is given as:

```
border-width : [ thin | medium | thick | length ]
1 to 4 values | inherit
```

Example:

```
<!DOCTYPE html>
<html>
<head>
<style>

.demo_container{
  width:500px;
  margin:0 auto;
  text-align: center;
  justify-content: center;
  align-items: center;
}

p.one {
  border-color: rgb(221, 104, 104);
  border-style: solid;
      border-width: 2px;
  }
  p.two {
    border-color: rgb(193, 221, 90);
  border-style: solid;
      border-width: 2px 5px;
  }
  p.three {
    border-color: rgb(93, 231, 109);
  border-style: solid;
      border-width: 2px 5px 7px;
  }
  p.four {
    border-color: rgb(212, 106, 241);
  border-style: solid;
```

```
        border-width: medium 5px thick 5px;
    }
</style>
</head>
<body>
<div class="demo_container">
  <h1> CSS Border Property- Width </h1>
  <p class="one"> <strong> Having one-value
syntax:</strong> Here is the single value sets
the width of four border sides.</p>
  <p class="two"> <strong> Having two-value
syntax:</strong> the first value sets as  the
width of the top & bottom border, while the
second value sets the width the right and left
sides border.</p>
  <p class="three"> <strong> Having three-value
syntax:</strong> the first value as sets the
width of the top border, the second value sets
the width of right and left border, the third
value sets width of the the bottom border.</p>
  <p class="four"> <strong> Here having four-
value syntax:</strong> each value sets the width
of the border individually in the order top,
right, bottom, and left.</p></div>
</body>
</html>
```

The output of the code is given below:

CSS Border Property- Width

Having one-value syntax: the single value sets the width of all four border sides.

Having two-value syntax: the first value sets the width of the top and bottom border, while the second value sets the width of the right and left sides border.

Having three-value syntax: the first value sets the width of the top border, the second value sets the width of the right and left border, and the third value sets the width of the the bottom border.

Having four-value syntax: each value sets the width of the border individually in the order top, right, bottom, and left.

CSS border property-width.

2. border-bottom property: The border-bottom CSS property sets the width, style, and color of the bottom border of an element. It is a property for setting the individual bottom border properties, that is, border-bottom-width, border-bottom-style, and border-bottom-color at once. The syntax of this property is given as:

```
border-bottom: [ border-bottom-width border-
bottom-style border-bottom-color ] | initial |
inherit
```

Example:

```
<!DOCTYPE html>
<html>
<head>
<style>

.demo_container{
  width:400px;
  margin:0 auto;
  text-align: center;
  justify-content: center;
  align-items: center;
}

.text-1 {
          border-left: 4px solid #ee2525;

   }
   .text-2{
     border-right: 3px solid #26e126;
   }
     .text-3{
            border-top: 4px solid #d42323;
   }
     .text-4{
     border-bottom: 3px solid #24d024;
   }
```

```
</style>
</head>
<body>
<div class="demo_container">
  <h1> CSS Border Property - Left, Bottom,
Right, Top </h1>
  <p class="text-1">Lorem dolor sit amet
consectetur adipisicing elit. Maxime mollitia,
molestiae quas vel sint commodi repudiandae
consequuntur voluptatum laborum numquam
blanditiis harum quisquam eius sed odit fugiat
iusto fuga praesentium optio, eaque rerum!</h1>
<br>
  <br>
  <p class="text-2"> Provident similique
accusantium nemo autem. Veritatis obcaecati
tenetur iure eius earum ut molestias architecto
voluptate aliquam nhil, eveniet aliquid culpa
officia aut! .</p> <br>
  <br>
  <p class="text-3">Lorem ipsum dolor sit amet
consectetur adipisicing elit. Maxime mollitia,
molestiae quas vel sint commodi repudiandae
consequuntur voluptatum laborum numquam
blanditiis harum quisquam eius sed odit fugiat
iusto fuga praesentium optio, eaque rerum!</h1>
<br>
  <br>
  <p class="text-4"> Provident similique
accusantium nemo autem. Veritatis obcaecati
tenetur iure eius earum ut molestias architecto
voluptate aliquam nihil, eveniet aliquid culpa
officia aut! .</p><br>
  </body>
</html>
```

The output of the code is given below:

CSS Border Property - Left, Bottom, Right, Top

Lorem ipsum dolor sit amet consectetur adipisicing elit. Maxime mollitia, molestiae quas vel sint commodi repudiandae consequuntur voluptatum laborum numquam blanditiis harum quisquam eius sed odit fugiat iusto fuga praesentium optio, eaque rerum!

Provident similique accusantium nemo autem. Veritatis obcaecati tenetur iure eius earum ut molestias architecto voluptate aliquam nhil, eveniet aliquid culpa officia aut! .

Lorem ipsum dolor sit amet consectetur adipisicing elit. Maxime mollitia, molestiae quas vel sint commodi repudiandae consequuntur voluptatum laborum numquam blanditiis harum quisquam eius sed odit fugiat iusto fuga praesentium optio, eaque rerum!

Provident similique accusantium nemo autem. Veritatis obcaecati tenetur iure eius earum ut molestias architecto voluptate aliquam nihil, eveniet aliquid culpa officia aut! .

CSS border property – left, bottom, right, top.

3. border-bottom-color property: The border-bottom color property sets the color of an element's bottom border individually. However, in other cases the shorthand properties like border color or border bottom are more convenient to use and preferable. The syntax of this property is given as:

```
border-bottom-color:  color | transparent |
inherit
```

Example:

```html
<!DOCTYPE html>
<html>
<head>
<style>

.demo_container{
  width:400px;
  margin:0 auto;
  text-align: center;
  justify-content: center;
  align-items: center;
}

.text-1 {
  border-style: solid;
      border-bottom-color: #d83838;

  }
  .text-2{
    border-style: solid;
      border-left-color: #aec730;
  }
    .text-3{
      border-style: solid;
      border-right-color: #e12424;
  }
    .text-4{
      border-style: solid;
      border-top-color: #ffaa00;
  }
</style>
</head>
<body>
<div class="demo_container">
  <h1> CSS Border Bottom Property - Left, Bottom,
Right, Top </h1>
  <p class="text-1">Lorem ipsum dolor sit amet
consectetur adipisicing elit. Maxime mollitia,
molestiae quas vel sint commodi repudiandae
consequuntur voluptatum laborum numquam blanditiis
```

```
harum quisquam eius sed odit fugiat iusto fuga
praesentium optio, eaque rerum!</h1> <br>
  <br>
  <p class="text-2"> Provident similique
accusantium nemo autem. Veritatis obcaecati
tenetur iure eius earum ut molestias architecto
voluptate aliquam nhil, eveniet aliquid culpa
officia aut! .</p> <br>
  <br>
  <p class="text-3">Lorem ipsum dolor sit amet
consectetur adipisicing elit. Maxime mollitia,
molestiae quas vel sint commodi repudiandae
consequuntur voluptatum laborum numquam blanditiis
harum quisquam eius sed odit fugiat iusto fuga
praesentium optio, eaque rerum!</h1> <br>
  <br>
  <p class="text-4"> Provident similique
accusantium nemo autem. Veritatis obcaecati
tenetur iure eius earum ut molestias architecto
voluptate aliquam nihil, eveniet aliquid culpa
officia aut! .</p><br>
  </body>
</html>
```

4. border-bottom-left-radius property: The border-bottom-left-radius property sets the rounded shape for the "bottom-left" corner of an element border box.

Example:

```
<!DOCTYPE html>
<html>
<head>
<style>

.demo_container{
  width:400px;
  margin:0 auto;
```

```
  text-align: center;
  justify-content: center;
  align-items: center;
}

.text-1 {
  border-style: solid;
  border-bottom-left-radius: 20px;

    }
    .text-2{
      border-style: solid;
    border-bottom-right-radius: 20px;
    }

</style>
</head>
<body>
<div class="demo_container">
  <h1> CSS Border Bottom Left Property - Left,
Right </h1>
  <p class="text-1"> Lorem sit amet consectetur
adipisicing elit. Maxime mollitia, molestiae
quas vel sint commodi repudiandae consequuntur
voluptatum laborum numquam blanditiis harum
quisquam eius sed odit fugiat iusto fuga
praesentium optio, eaque rerum!</h1> <br>
  <br>
  <p class="text-2"> Provident similique
accusantium nemo autem. Veritatis obcaecati
tenetur iure eius earum ut molestias architecto
voluptate aliquam nhil, eveniet aliquid culpa
officia aut! .</p> <br>
  <br>
  </body>
</html>
```

The output of the code is given below:

CSS Border Bottom Property - Left, Bottom, Right, Top

Lorem ipsum dolor sit amet consectetur adipisicing elit. Maxime mollitia, molestiae quas vel sint commodi repudiandae consequuntur voluptatum laborum numquam blanditiis harum quisquam eius sed odit fugiat iusto fuga praesentium optio, eaque rerum!

Provident similique accusantium nemo autem. Veritatis obcaecati tenetur iure eius earum ut molestias architecto voluptate aliquam nhil, eveniet aliquid culpa officia aut! .

Lorem ipsum dolor sit amet consectetur adipisicing elit. Maxime mollitia, molestiae quas vel sint commodi repudiandae consequuntur voluptatum laborum numquam blanditiis harum quisquam eius sed odit fugiat iusto fuga praesentium optio, eaque rerum!

Provident similique accusantium nemo autem. Veritatis obcaecati tenetur iure eius earum ut molestias architecto voluptate aliquam nihil, eveniet aliquid culpa officia aut! .

CSS border bottom left property – left, right.

5. border-bottom-style property: The border-bottom style property sets the style of an element's bottom border individually. However, in other cases the shorthand properties like border style or border bottom are more convenient to use and preferable.

The syntax of this property is given as:

```
border-bottom-style: none | hidden | dashed |
dotted |solid | double | groove | ridge | inset |
outset | initial | inherit.
```

Example:

```
<!DOCTYPE html>
<html>
<head>
<style>

.demo_container{
  width:400px;
  margin:0 auto;
  text-align: center;
  justify-content: center;
  align-items: center;
}


    .none {
        border-bottom-style: none;
    }
    .dotted {
        border-bottom-style: dotted;
    }
    .dashed {
        border-bottom-style: dashed;
    }
    .solid {
        border-bottom-style: solid;
    }
    .double {
        border-bottom-style: double;
    }
    .groove {
        border-bottom-style: groove;
    }
    .ridge {
        border-bottom-style: ridge;
    }
    .inset {
        border-bottom-style: inset;
    }
    .outset {
        border-bottom-style: outset;
    }
```

```
    </style>
  </style>
  </head>
  <body>
  <div class="demo_container">
    <h1> CSS Various Border style  </h1>
    <p class="none">Lorem dolor sit amet
consectetur adipisicing elit.esentium optio,
eaque rerum!</h1> <br>
    <br>
    <p class="dotted">Lorem ipsum dolor sit amet
consectetur adipisicing elit.esentium optio,
eaque rerum!</h1> <br>
    <br>
    <p class="none">Lorem dolor sit amet
consectetur adipisicing elit.esentium optio,
eaque rerum!</h1> <br>
    <br>
    <p class="dashed">Lorem ipsum dolor sit amet
consectetur adipisicing elit.esentium optio,
eaque rerum!</h1> <br>
    <br>
    <p class="solid">Lorem ipsum dolor sit amet
consectetur adipisicing elit.esentium optio,
eaque rerum!</h1> <br>
    <br>
    <p class="double">Lorem ipsum dolor sit amet
consectetur adipisicing elit.esentium optio,
eaque rerum!</h1> <br>
    <br>
    <p class="groove">Lorem ipsum dolor sit amet
consectetur adipisicing elit.esentium optio,
eaque rerum!</h1> <br>
    <br>
    <p class="ridge">Lorem ipsum dolor sit amet
consectetur adipisicing elit.esentium optio,
eaque rerum!</h1> <br>
    <br>
    <p class="inset">Lorem ipsum dolor sit amet
consectetur adipisicing elit.esentium optio,
eaque rerum!</h1> <br>
    <br>
```

```
<p class="outset">Lorem ipsum dolor sit amet
consectetur adipisicing elit.esentium optio,
eaque rerum!</h1> <br>
  <br>


  </body>
</html>
```

The output of the code is given below:

CSS Various Border style

Lorem ipsum dolor sit amet consectetur adipisicing
elit.esentium optio, eaque rerum!

Lorem ipsum dolor sit amet consectetur adipisicing
elit.esentium optio, eaque rerum!

Lorem ipsum dolor sit amet consectetur adipisicing
elit.esentium optio, eaque rerum!

Lorem ipsum dolor sit amet consectetur adipisicing
elit.esentium optio, eaque rerum!

Lorem ipsum dolor sit amet consectetur adipisicing
elit.esentium optio, eaque rerum!

Lorem ipsum dolor sit amet consectetur adipisicing
elit.esentium optio, eaque rerum!

Lorem ipsum dolor sit amet consectetur adipisicing
elit.esentium optio, eaque rerum!

Lorem ipsum dolor sit amet consectetur adipisicing
elit.esentium optio, eaque rerum!

Lorem ipsum dolor sit amet consectetur adipisicing
elit.esentium optio, eaque rerum!

Lorem ipsum dolor sit amet consectetur adipisicing
elit.esentium optio, eaque rerum!

CSS various border style.

6. border-image property: The border-image CSS property specifies how an image is to be used in place of the border styles. This
is a property for setting border-image-source, border-image-
width, border-image-slice, border-image-outset, and border-
image-repeat properties at once. The syntax of this property is
given as:

```
border-image: [ source slice width outset repeat ]
| initial | inherit
```

Example:

```
<!DOCTYPE html>
<html>
<head>
<style>

.demo_container{
  width:400px;
  margin:0 auto;
  text-align: center;
  justify-content: center;
  align-items: center;
}
.box {
    width: 300px;
    height: 150px;
    border: 15px solid transparent;
    border-image: url("/images-1.jpg") 30 30
round;
  }
    </style>
</style>
</head>
<body>
<div class="demo_container">
  <h1> CSS border-image Property  </h1>
  <div class="box"></div>

  </body>
</html>
```

The output of the code is given below:

CSS border-image Property

CSS border-image property.

CSS COLOR PROPERTIES

The CSS color feature sets the color of the text element content. All modern web browsers support different colors to maintain a colorful display. In CSS, colors can be represented in several ways and even use a color word such as "color: blue." However, this method only supports names for certain colors. Therefore, in CSS, other advanced methods are used to display colors such as RGB, HSL, HEX, etc.

Example:

```
<!DOCTYPE html>
<html>
<head>
<style>

.demo_container{
  width:400px;
  margin:0 auto;
  text-align: center;
  justify-content: center;
  align-items: center;
  color:red
}

    .c1{
      color: black;
    }
    .c2 {
```

```
            color:green
            }

        </style>
</style>
</head>
<body>
<div class="demo_container">
    <h1> CSS Color Propterty  </h1>
    <p class="c1">Here we use color with Simple
color code (Black). </h1> <br>
        <br>
        <p class="c2"> Here we use color with Simple
color code (Green). </h1> <br>
        <br>
    </body>
</html>
```

The output of the code is given below:

CSS Color Propterty

Here we use color with Simple color code (Black).

Here we use color with Simple color code (Green).

CSS color property (color).

RGB Colors

RGB is a combination of three colors (red, green, and blue) which are used in all computer programs to display colors. As we know, these are the basic colors, and by combining them, we can find any color that appears in the color spectrum.

Example:

```
<!DOCTYPE html>
<html>
<head>
<style>

.demo_container{
    width:400px;
```

```
    margin:0 auto;
    text-align: center;
    justify-content: center;
    align-items: center;
}

    .c1{
      color: rgb(205, 92, 92) ;
    }
    .c2 {
      color:rgb(100, 149, 237)
      }

    </style>
</style>
</head>
<body>
<div class="demo_container">
  <h1> CSS Color Property (RGB) </h1>
  <p class="c1">Here we use color with RGB color
code (rgb(205, 92, 92)). </h1> <br>
    <br>
    <p class="c2"> Here we use color with RGB
color code (rgb(100, 149, 237)). </h1> <br>
    <br>
  </body>
</html>
```

The output of the code is given below:

CSS Color Property (RGB)

Here we use color with RGB color code (rgb(205, 92, 92)).

Here we use color with RGB color code (rgb(100, 149, 237)).

CSS color property (RGB).

RGBA Colors

In CSS, RGBA is also a color display format with Alpha extension. The structure of this color work is given below:

```
rgba (Red, Green, Blue, Alpha)
```

In this work, Alpha is used to express color blurring. In CSS opacity, the area is used to set the color brightness and its width is between 0.0 and 1.0, where 0.0 stands for absolute transparency and 1.0 represents absolute opaque. You will understand better in the example provided.

```
<!DOCTYPE html>
<html>
<head>
<style>

.demo_container{
  width:400px;
  margin:0 auto;
  text-align: center;
  justify-content: center;
  align-items: center;
}

    .c1{
      background-color: rgba(11, 156, 49, 0.2)  ;
    }
    .c2 {
      background-color: RGBA(11, 156, 49, 0.4)  ;
      }
    .c3{
      background-color: rgba(11, 156, 49, 0.6)  ;
    }
    .c4 {
      background-color: RGBA(11, 156, 49, 0.8)  ;
      }

    .c5{
      background-color: rgba(11, 156, 49, 1)  ;
    }


    </style>
</style>
</head>
<body>
<div class="demo_container">
  <h1> CSS Color Property (RGBA) </h1>
```

```
<p class="c1">Here we use color with RGBA color code
(rgba(11, 156, 49, 0.2). </h1> <br>
    <br>
    <p class="c2"> Here we use color with RGBA color
code (RGBA(11, 156, 49, 0.4)). </h1> <br>
    <br>
    <p class="c3">Here we use color with RGBA color
code (rgba(11, 156, 49, 0.6)). </h1> <br>
    <br>
    <p class="c4"> Here we use color with RGBA color
code RGBA(11, 156, 49, 0.8)  ). </h1> <br>
    <br>
    <p class="c5">Here we use color with RGBA color
code (rgba(11, 156, 49, 1)). </h1> <br>
    </body>
</html>
```

The output of the code is given below:

CSS Color Property (RGBA)

Here we use color with RGBA color code (rgba(11, 156, 49, 0.2).

Here we use color with RGBA color code (RGBA(11, 156, 49, 0.4)).

Here we use color with RGBA color code (rgba(11, 156, 49, 0.6)).

Here we use color with RGBA color code RGBA(11, 156, 49, 0.8)).

Here we use color with RGBA color code (rgba(11, 156, 49, 1)).

CSS color property (RGBA).

CSS HEX Colors

CSS colors can also be defined by hexadecimal values, which is another way of representing colors. In CSS, it is the most common way to define color using hexadecimal values with a "#" symbol such as #RRGGBB. Although, R, G, B codes red, green, and blue, respectively.

Hexadecimal numbers with a combination of 0-9 and A-F are used to represent color in CSS. Some examples of HEX basic colors are given below.

```
<!DOCTYPE html>
<html>
<head>
<style>

.demo_container{
  width:400px;
  margin:0 auto;
  text-align: center;
  justify-content: center;
  align-items: center;
  color:red
}

    .c1{
      color: #6495ed   ;
    }
    .c2 {
      color:#b22222
      }

    </style>
</style>
</head>
<body>
<div class="demo_container">
  <h1> CSS Color Property (HEX) </h1>
  <p class="c1">Here we use color with HEX color code
(#6495ed). </h1> <br>
    <br>
    <p class="c2"> Here we use color with HEX color
code (#b22222). </h1> <br>
    <br>
  </body>
</html>
```

The output of the code given below:

CSS Color Property (HEX)

Here we use color with HEX color code (#6495ed).

Here we use color with HEX color code (#b22222).

CSS color property (HEX).

1. opacity property: The opacity property specifies the transparency of an element. The syntax of this property is given as:

```
opacity:   alphavalue | initial | inherit
```

Example:

```
<!DOCTYPE html>
<html>
<head>
<style>

.demo_container{
  width:400px;
  margin:0 auto;
  text-align: center;
  justify-content: center;
  align-items: center;
}
.c1{
  opacity: 0.4;
}
img{
  background-image: url('/images-1.jpg');
  width: "300px";
  height: "300px";
  opacity: 0.5;

  }

  </style>
</style>
</head>
```

```
<body>
<div class="demo_container">
  <h1> CSS Color Opacity Property </h1>
  <h2 class="c1">Here we use color with Opacity
code. </h2> <br>
    <img src="/images-1.jpg" alt="image"
width="200px" height="200px">
  </body>
</html>
```

2. The output of the code is given below:

CSS Color Opacity Property

Here we use color with Opacity code.

CSS color opacity property.

CSS DIMENSION PROPERTIES

1. height property: The height CSS property specifies the height of the content area of an element. The content area does not include padding, borders, or margins.

The syntax of this property is given as:

```
height:   length | percentage | auto | initial |
inherit
```

Example:

```html
<!DOCTYPE html>
<html>
<head>
<style>

.demo_container{
  width:300px;
  margin:0 auto;
  text-align: center;
  justify-content: center;
  align-items: center;
}
.c1{
  background-color: palevioletred;
  height: 200px;
  width:300px;

}
img{
  background-image: url('/images-1.jpg');
  width: 300px;
   height: 300px;


    }

    </style>
</style>
</head>
<body>
<div class="demo_container">
  <h1> CSS Dimension Property - Height,  Width </h1>
  <h2 class="c1">Having height "200" and width "300"  </h2> <br>
    <img src="/images-1.jpg" alt="image" width="200px" height="200px">
  </body>
</html>
```

The output of the code is given below:

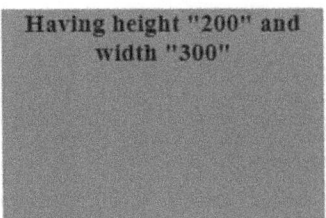

CSS dimension property – height, width.

2. max-height and max-width property: The max-height CSS property specifies the maximum height of the content area of an element. This maximum height does not include padding, borders, or margins.

The max-width property defines the max width of the content area of an element. This maximum width does not include padding, borders, or margins

The syntax of this property is given as:

```
max-height:  length | percentage | none | initial
| inherit
```

The syntax of this property is given as:

```
max-width:  length | percentage | none | initial |
inherit
```

Example:

```
<!DOCTYPE html>
<html>
<head>
<style>

.demo_container{
  width:500px;
  height:500px;
  margin:0 auto;
  text-align: center;
  justify-content: center;
  align-items: center;
}
.c1{
  background-color: palevioletred;
  max-height: 300px;
  max-width:300px;

}
img{
  background-image: url('/images-1.jpg');
  max-height: 200px;
  max-width:300px;

    }

    </style>
</style>
</head>
<body>
<div class="demo_container">
  <h1> CSS Dimension Property - Height,  Width
</h1>
  <p class="c1">Having max-height "200" and
max-width "300"  </p> <br>
    <img src="/images-1.jpg" alt="image"
width="200px" height="200px">
  </body>
</html>
```

The output of the code is given below:

CSS Dimension Property - Max-Height , Max-Width

Having max-height "200" and max-width "300"

CSS dimension property – max-height, max-width.

3. min-height property: The min-height CSS property specifies the minimum height of the content area of an element. This minimum height does not include padding, borders, or margins.

The syntax of this property is given as:

```
min-height:    length | percentage | initial |
inherit
```

Example:

```
<!DOCTYPE html>
<html>
<head>
<style>

.demo_container{
  width:800px;
  height:500px;
  margin:0 auto;
  text-align: center;
  justify-content: center;
  align-items: center;
}
.c1{
  background-color: palevioletred;
  min-height: 200px;
  min-width:200px;

}
```

```
img{
  background-image: url('/images-1.jpg');
  min-height: 200px;
  min-width:200px;

    }

    </style>
</style>
</head>
<body>
<div class="demo_container">
  <h1> CSS Dimension Property - Min-Height,
Min-Width </h1>
  <p class="c1">Having min-height "200" and
min-width "300"   </p> <br>
    <img src="/images-1.jpg" alt="image"
width="200px" height="200px">
  </body>
</html>
```

CSS GENERATED CONTENT PROPERTIES

1. content property: The content CSS property is used in combination
   with the ::before and ::after pseudo-elements to generate content in
   an element. The syntax of this property is given as:

   ```
   content:  normal | none | counter | string |
   url(url) | attr(attribute) | open-quote | close-
   quote | no-open-quote | no-close-quote | initial |
   inherit
   ```

   Example:

   ```
   <!DOCTYPE html>
   <html>
   <head>
   <style>

   .demo_container{
     width:800px;
   ```

```
    height:500px;
    margin:0 auto;
    text-align: center;
    justify-content: center;
    align-items: center;
}
.before::before{
    content: "Hi! ";
    display: inline;
    background-color: rgb(211, 81, 81);
}

.after::after{
    content: "Bye! ";
    display: inline;
    background-color: rgb(224, 79, 79);

}

    </style>
</style>
</head>
<body>
<div class="demo_container">
    <h1> CSS Content Property </h1>
    <p class="before"> The text will be added
before every paragraph in your code </p>
    <p class="after"> The text will be added
before every paragraph in your code </p>

</body>
</html>
```

The output of the code is given below:

CSS Content Property

Hi! The text will be added before every paragraph in your code

The text will be added before every paragraph in your code Bye!

CSS content property.

2. quotes property: The quotes CSS property specifies the quotation marks for the embedded quotations. The quotes characters specified for this property are used for the open-quote and close-quote values of the content property.

The syntax of this property is given as:

```
quotes:    [string string]one or more pairs | none
| initial | inherit
```

Example:

```
<!DOCTYPE html>
<html>
<head>
<style>

.demo_container{
  width:800px;
  height:500px;
  margin:0 auto;
  text-align: center;
  justify-content: center;
  align-items: center;
}
    q{
        quotes: '[' ']' '"' '"';
        }
        .q::before {
            content: open-quote;
        }
        .q::after {
            content: close-quote;
        }
    </style>
</style>
</head>
<body>
<div class="demo_container">
  <h1> CSS Content Property </h1>
  <p class="before"> <q> The text will be added
before every paragraph in your code with <q>
embedded qutations</q></q> </p>
```

```
<p class="after"> The text will be added
before every paragraph in your code  <q>
embedded qutations </q> </p>

</body>
</html>
```

3. counter-reset property: The counter-reset CSS property is used in combination with the counter-increment property for creating auto-incrementing counters, and with the content property to display the generated counter values.

The syntax of the property is given as:

```
counter-reset: [ identifier integer ]1 or more
pairs | none | initial | inherit
```

Example:

```
<!DOCTYPE html>
<html>
<head>
<style>

.demo_container{
  width:800px;
  height:500px;
  margin:0 auto;
  text-align: center;
  justify-content: center;
  align-items: center;
}
h1 {
    counter-reset: category;
  }
  .before::before {
    counter-increment: section;
    content: "Section " counter(section) ". ";
  }
  .after::before {
    counter-increment: category;
    content: counter(section) "."
counter(category) " ";
  }
```

```
    li{
       list-style-type: none;
    }
    </style>
</style>
</head>
<body>
<div class="demo_container">
   <h1> CSS Counter Reset Property </h1>
   <ul class="before">
      <li>A</li>
   </ul>
   <ul class="after">
      <li>B</li>
   </ul>

</body>
</html>
```

The output of the code given below:

CSS Counter Reset Property

Section 1.
A

0.1
B

CSS counter-reset property.

CSS FLEXIBLE BOX LAYOUT

The alignment-content feature changes the behavior of the flex-wrap structure. Aligns dynamic lines. Used to specify alignment between lines within a flexible container. This feature describes how each flexible line is aligned within the flexbox and only works if the flex-wrap: wrap is used, that is, when multiple flexbox lines are present.

List of content alignment items:

1. center : In this, the items are positioned at the center of the flex container.

2. stretch : In this, the items are stretched to fit the flex container. The free-space is divided between all the items. This is the default value.

3. flex-start : In this, the items are positioned at the beginning of the flex container.

4. flex-end : end   In this, the items are positioned at the end of the flex container.

5. space-around : In this, the items are evenly distributed in the flex container so that the spaces around (i.e., before, between, and after) every item are the same.

6. space-between: In this, the items are evenly distributed in the flex container in such a way that the space between two adjacent items is the same.

7. initial : It sets property to its default value.

8. Inherit : If specified, the element takes the calculated value of its parent element's align-content property.

The syntax of this property is given as:

```
align-content: center | flex-start | flex-end | space-
between | space-around | stretch | initial | inherit
```

Example:

```
<!DOCTYPE html>
<html>
<head>
<style>

.flex-container {
  display: flex;

  flex-flow:  wrap;

  justify-content: center;

  padding: 0;
  margin: 0;
  list-style: none;
}
```

```
.flex-item {
  background: tomato;
  padding: 5px;
  width: 200px;
  height: 150px;
  margin-top: 10px;
  line-height: 150px;
  color: white;
  font-weight: bold;
  font-size: 3em;
  text-align: center;
}
</style>
</head>
<body>
<h1> CSS Flexbox Layout </h1>
<ul class="flex-container">
  <li class="flex-item">1</li>
  <li class="flex-item">2</li>
  <li class="flex-item">3</li>
  <li class="flex-item">4</li>
  <li class="flex-item">5</li>
  <li class="flex-item">6</li>
</ul>

</body>
</html>
```

The output of the code is given below:

CSS Flexbox Layout

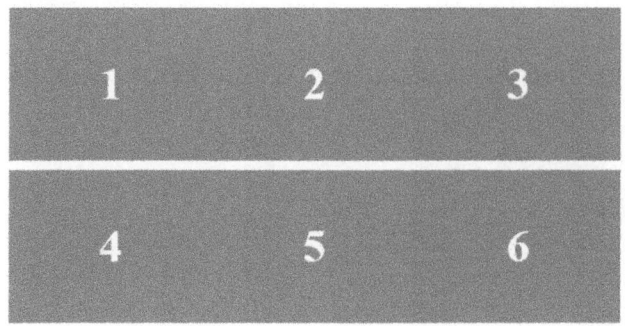

CSS Flexible Layout.

CSS FONT PROPERTIES

The font property sets the style, variant, boldness, line-height, and the font family for an element's text content. It is a property for setting the individual font properties, that is, font-style, font-variant, font-weight, font-size, line-height, and font-family in a single declaration.

The syntax of the property is given as:

```
font:   [ font-style font-variant line-height font-
family ] | caption | icon | menu | message-box |
small-caption | status-bar | initial | inherit
```

Example:

```
<!DOCTYPE html>
<html>
<head>
<style>

.demo_container{
  width:800px;
  height:500px;
  margin:0 auto;
  text-align: center;
  justify-content: center;
  align-items: center;
}
.before{
  font: bold 2.5em "Times New Roman", Times,
serif;
}

.after{
  font: normal 1.2em Arial, Helvetica, sans-serif;

}

    </style>
</style>
</head>
<body>
<div class="demo_container">
  <h1> CSS Font Property </h1>
```

```
    <p class="before"> The text will be in font
(  font: bold 2.5em "Times New Roman", Times,
serif;  ) </p>
    <p class="after"> The text will be in font
(   font: normal 1.2em Arial, Helvetica, sans-
serif;   ) </p>

</body>
</html>
```

The output of the code is given below:

CSS Font Style Property

The text will be in font (font-style: italic;)

The text will be in font (font-style:normal;)

CSS font style property.

1. font-size property: The font-size CSS property sets the font size for the element's text content. The syntax of the property is given as:

```
font-size:  xx-small | x-small | small | medium |
large | x-large | xx-large | smaller | larger |
length | percentage | initial | inherit
```

Example:

```
<!DOCTYPE html>
<html>
<head>
<style>

.demo_container{
  width:400px;
  margin:0 auto;
  text-align: center;
  justify-content: center;
  align-items: center;
}
.a{
```

```
    font-size: xx-small
}

.b{
font-size: x-small
}
.c{
    font-size: small
}

.d{
font-size: medium
}
.e{
    font-size: x-large
}

.f{
font-size: xx-large
}
.g{
font-size: smaller
}
.h{
font-size: larger
}
</style>
</head>
<body>
<div class="demo_container">
  <h1> CSS Font Size Property </h1>
  <p class="a"> The text will be in font
(   font-size: xx-small    ) </p>
  <p class="b"> The text will be in font
(   font-size: x-small    ) </p>
  <p class="c"> The text will be in font
(   font-size: small    ) </p>
  <p class="d"> The text will be in font
(   font-size: xx-small    ) </p>
  <p class="e"> The text will be in font
(    font-size: x-large    ) </p>
  <p class="f"> The text will be in font
(   font-size: xx-large    ) </p>
```

```
<p class="g"> The text will be in font
(   font-size: smaller     ) </p>
   <p class="h"> The text will be in font
(  font-size: larger ) </p>

</body>
</html>
```

The output of the code is given below:

CSS Font Size Property

The text will be in font (font-size: xx-small)

The text will be in font (font-size: x-small)

The text will be in font (font-size: small)

The text will be in font (font-size: xx-small)

The text will be in font (font-size: x-large)

The text will be in font (font-size: xx-large)

The text will be in font (font-size: smaller)

The text will be in font (font-size: larger)

CSS font size property.

2. font-size-adjust property: The CSS feature of the font size adjustment specifies that the font size should be selected based on lowercase characters than uppercase.

If the original font family selection mentioned by the author is not available when font back occurs, that may result in a large or small font size. The font size change feature is a way to keep text readings in such a state. It does this by adjusting the font size so that the x length is the same regardless of the font used.

CSS LIST PROPERTIES

1. list-style property: The list-style CSS property defines the display style for list items. It is a shorthand property for setting the individual list properties, that is, list-style-type, list-style-position, and list-style-image in a single declaration.

   The syntax of the property is given as:

```
list-style: [ list-style-type list-style-position
list-style-image ] | initial | inherit
```

Example:

```
<!DOCTYPE html>
<html>
<head>
<style>

.demo_container{
  width:400px;
  margin:0 auto;
  text-align: center;
  justify-content: center;
  align-items: center;
}
ul {
    list-style: circle inside;
  }
  ol {
    list-style: upper-latin outside;
  }

</style>
</head>
<body>
<div class="demo_container">
  <h1> CSS list-style Property </h1>
  <h2> Unordered List (Inside) </h2>
  <ul>
    <li> Item 1 </li>
    <li> Item 2 </li>
    <li> Item 3 </li>
  </ul>
```

```
<h2> Ordered List (Outside) </h2>
<ol>
    <li> Item 1 </li>
    <li> Item 2 </li>
    <li> Item 3 </li>
</ol>
</body>
</html>
```

The output of the code is given below:

Unordered List (Inside)

- ○ Item 1
- ○ Item 2
- ○ Item 3

Ordered List (Outside)

A.	Item 1
B.	Item 2
C.	Item 3

CSS list-style property.

2. list-style-type property: The list-style-type CSS property specifies the type of marker for the list-items.

```
list-style-type:  disc | circle | square |
decimal| lower-roman | upper-roman | lower-greek |
lower-latin | upper-latin | georgian | lower-alpha
| upper-alpha | none | initial | inherit
```

Example:

```
<!DOCTYPE html>
<html>
<head>
<style>

.demo_container{
  width:400px;
  margin:0 auto;
}
```

```
ol {
      list-style-type: decimal-leading-zero
inside;
    }
    ul {
       list-style-type: square;
    }
</style>
</head>
<body>
<div class="demo_container">
  <h2> Unordered List ( decimal-leading-zero
inside ) </h2>
  <ul>
      <li> Item 1 </li>
      <li> Item 2 </li>
      <li> Item 3 </li>
  </ul>
  <h2> Ordered List ( square ) </h2>
  <ol>
      <li> Item 1 </li>
      <li> Item 2 </li>
      <li> Item 3 </li>
  </ol>
</body>
</html>
```

The output of the code is given below:

CSS list-style-type Property

Unordered List (decimal-leading-zero inside)

- Item 1
- Item 2
- Item 3

Ordered List (square)

1. Item 1
2. Item 2
3. Item 3

CSS list-style-type property.

Here you will get various values of list-style-type property.

Value	Description
disc	The marker is a filled circle.
circle	The marker is a filled hollow circle.
square	
decimal	The marker is as a decimal number Beginning with 1.
decimal-leading-zero	The marker is as a decimal number Padded by initial zero For example, 01, 02, 03, ... 10, 11
lower-greek	The marker is a lowercase Greek letters alpha, beta, gamma, ... For example, α, β, γ, ...
upper-roman	The marker is as uppercase Roman numerals For example, I, II, III, IV, V, ...
lower-roman	The marker is as lowercase Roman numerals For example, i, ii, iii, iv, v, ...
lower-greek	The marker is as lowercase Greek letters alpha, beta, gamma, ... For example, α, β, γ, ...
lower-Latin	The marker is as lowercase Latin letters For example, a, b, c, ... z
upper-Latin	The marker is as uppercase Latin letters For example, A, B, C, ... Z
lower-alpha	The marker is as uppercase Latin letters For example, a, b, c, ... z
upper-alpha	The marker is as uppercase Latin letters For example, A, B, C, ... Z
Armenian	The marker is as traditional Armenian numbering such as For example, ayb/ayp, ben/pen, gim/keem, ...
Georgian	The marker is as traditional of the Georgian numbering For example, an, ban, gan, ... he, tan, in ...

3. **list-style-image property:** The list-style-image property specifies an image to be used as a list-item marker.

The syntax of the property is given as:

```
such as  URL | none | initial | inherit
```

Example:

```
<!DOCTYPE html>
<html>
<head>
<style>
```

```
.demo_container{
  width:400px;
  margin:0 auto;
}
ul {
     list-style-image: url("/images-1.jpg");


    }

</style>
</head>
<body>
<div class="demo_container">
  <h1> CSS list-style-image Property </h1>
  <h2> Unordered List </h2>
  <ul>
     <li> Item 1 </li>
     <li> Item 2 </li>
     <li> Item 3 </li>
  </ul>
</body>
</html>
```

4. list-style-position property: The list-style-position CSS property defines the position of the list-item marker with respect to the list item's block box. The syntax of this property is given as:

```
list-style-position: inside | outside | initial |
inherit
```

Example:

```
<!DOCTYPE html>
<html>
<head>
<style>

.demo_container{
  width:400px;
  margin:0 auto;
}
ol {
     list-style-position: inside;
    }
```

```
    ul {
        list-style-position: outside;
    }
  ol li, ul li{
        background: #d8bfd8;
    }
</style>
</head>
<body>
<div class="demo_container">
  <h1> CSS list-style-position Property </h1>
  <h2> List Marker Positioned Inside </h2>
  <ol>
      <li> Item 1 </li>
      <li> Item 2 </li>
      <li> Item 3 </li>
  </ol>
  <h2> List Marker Positioned Outside </h2>
  <ul>
      <li> Item 1 </li>
      <li> Item 2 </li>
      <li> Item 3 </li>
  </ul>
</body>
</html>
```

The output of the code is given below:

CSS list-style-position Property

List Marker Positioned Inside

1. Item 1
2. Item 2
3. Item 3

List Marker Positioned Outside

- Item 1
- Item 2
- Item 3

CSS list-style-position property.

CSS MARGIN PROPERTIES

The margin property sets the margin on all four sides of the element. It is a shorthand property for margin-top, margin-right, margin-bottom, and margin-left property.

The syntax of this property is given as:

```
margin:   [ length | percentage | auto ] 1 to 4 values
| initial | inherit
```

Example:

```
<!DOCTYPE html>
<html>
<head>
<style>

.demo_container{
  width:600px;
  margin:0 auto;
  text-align: center;
  justify-content: center;
  border:1px solid red ;
  align-items: center;
}

.text-1 {
  border-style: solid;
      margin: 25px;
    }
    .text-2{
      border-style: solid;
      margin: 20px 20px 20px 20px;
    }
      .text-3{
        border-style: solid;
        margin: 30px 30px;
    }
      .text-4{
        border-style: solid;
        margin: 30px 0px 30px 0px ;
    }
```

```
</style>
</head>
<body>
<div class="demo_container">
  <h1>CSS margin Property </h1>
  <p class="text-1"> Lorem ipsum sit amet
consectetur adipisicing elit. Maxime mollitia,
molestiae quas vel sint commodi repudiandae
consequuntur voluptatum laborum numquam blanditiis
harum quisquam eius sed odit fugiat iusto fuga
praesentium optio, eaque rerum!</h1> <br>
  <br>
  <p class="text-2"> Provident similique
accusantium nemo autem. Veritatis obcaecati
tenetur iure eius earum ut molestias architecto
voluptate aliquam nhil, eveniet aliquid culpa
officia aut! .</p> <br>
  <br>
  <p class="text-3">Lorem ipsum dolor sit
consectetur adipisicing elit. Maxime mollitia,
molestiae quas vel sint commodi repudiandae
consequuntur voluptatum laborum numquam blanditiis
harum quisquam eius sed odit fugiat iusto fuga
praesentium optio, eaque rerum!</h1> <br>
  <br>
  <p class="text-4"> Provident similique
accusantium autem. Veritatis obcaecati tenetur
iure earum ut molestias architecto voluptate
aliquam nihil, eveniet culpa officia aut!
.</p><br>
  </body>
</html>
```

This notation can take one, two, three, or four whitespace-separated values.

1. If set one value, then margin applies to all four sides.

2. If you set two values, then first value applies to the top and bottom, and the second value applies to the right and left side.

3. If set three values, then values apply to the top, horizontal (i.e., right and left) bottom side.

4. If set four values, then to the top, right, bottom, left side in the same order.

Instead of margin, you should use other various margin-top, margin-right, margin-bottom, and margin-left.

Example:

```
<!DOCTYPE html>
<html>
<head>
<style>

.demo_container{
  width:600px;
  margin:0 auto;
  text-align: center;
  justify-content: center;
  border: 2px solid black;
  align-items: center;
}

.text-1 {
  border:1px solid red ;
  margin-top: 25px;
    }
    .text-2{
      border:1px solid red ;
      margin-right: 20px ;
  }
    .text-3{
      border:1px solid red ;
      margin-bottom: 20px;
  }
    .text-4{
      border:1px solid red ;
```

```
        margin-left: 30px ;
    }
</style>
</head>
<body>
<div class="demo_container">
  <h1> CSS margin -  Property (margin top, margin
right, margin-bottom, margin-left) </h1>
  <p class="text-1">Lorem dolor sit amet
consectetur adipisicing elit. Maxime mollitia,
molestiae quas vel sint commodi repudiandae
consequuntur voluptatum laborum numquam blanditiis
harum quisquam eius sed odit fugiat iusto fuga
praesentium optio, eaque rerum!</h1> <br>
  <br>
  <p class="text-2"> Provident similique
accusantium nemo autem. obcaecati tenetur iure
eius earum ut molestias architecto voluptate
aliquam nhil, eveniet aliquid culpa officia aut!
.</p> <br>
  <br>
  <p class="text-3">Lorem ipsum dolor sit amet
consectetur adipisicing.  Maxime mollitia,
molestiae quas vel sint commodi repudiandae
consequuntur voluptatum laborum numquam blanditiis
harum quisquam eius sed odit fugiat iusto fuga
praesentium optio, eaque rerum!</h1> <br>
  <br>
  <p class="text-4"> Provident similique
accusantium nemo autem. Veritatis tenetur iure
eius earum ut molestias architecto voluptate
aliquam nihil, eveniet aliquid culpa officia aut!
.</p><br>
  </body>
</html>
```

The output of the code is given below:

CSS margin - Property (margin-top, margin-right, margin-bottom, margin-left)

Lorem ipsum dolor sit amet consectetur adipisicing elit. Maxime mollitia, molestiae quas vel sint commodi repudiandae consequuntur voluptatum laborum numquam blanditiis harum quisquam eius sed odit fugiat iusto fuga praesentium optio, eaque rerum!

Provident similique accusantium nemo autem. Veritatis obcaecati tenetur iure eius earum ut molestias architecto voluptate aliquam nhil, eveniet aliquid culpa officia aut! .

Lorem ipsum dolor sit amet consectetur adipisicing elit. Maxime mollitia, molestiae quas vel sint commodi repudiandae consequuntur voluptatum laborum numquam blanditiis harum quisquam eius sed odit fugiat iusto fuga praesentium optio, eaque rerum!

Provident similique accusantium nemo autem. Veritatis obcaecati tenetur iure eius earum ut molestias architecto voluptate aliquam nihil, eveniet aliquid culpa officia aut! .

CSS margin – property (margin-top, margin-right, margin-bottom, margin-left).

CSS MULTI-COLUMN LAYOUT PROPERTIES

1. CSS3 column-count property: The column count property defines the number of columns in a multi-column element.

   The syntax of this Property is given as:

   ```
   column-count:  number | auto | initial | inherit
   ```

   Example:

   ```
   <!DOCTYPE html>
   <html lang="en">
   <head>
   <meta charset="utf-8">
   <title>Example of CSS3 column-count Property
   </title>
   ```

```
<style>
  p {
        column-count: 2; /* Standard syntax */
  }
</style>
</head>
<body>
  <h1> CSS3 column-count Property </h1>
    <p>Lorem ipsum dolor amet, adipiscing elit.
```
Ut nisi egestas suscipit gravida. Sed velit
nisl, sed dui mollis, porta tempus ligula. vel
orci vel arcu pellentesque fermentum. Duis
bibendum metus arcu. Aliquam tortor vulputate,
sollicitudin felis a, mollis libero. Aliquam
consequat sapien, id blandit lectus. ac nibh ac
nulla tincidunt accumsan sit amet sit amet
risus. Integer id nisl urna. In a elementum,
auctor justo quis, tincidunt augue. Donec dui,
congue non neque quis, semper aliquam felis.
Praesent efficitur massa vel convallis euismod.
metus lectus, consectetur sit amet justo in,
venenatis faucibus nunc. Aenean faucibus, id
egestas convallis, felis mattis est, in
ultricies est urna ac nisl.

Cras placerat quis tortor quis. Nullam
imperdiet gravida velit eget sollicitudin.
dictum pretium justo vel congue. Praesent auctor
leo maximus aliquam, eget vehicula tortor.
Vestibulum finibus venenatis dui, nec lobortis
mauris convallis id. Maecenas porttitor erat,
at vulputate eros euismod a. In aliquam, dolor
et bibendum consequat, eros felis ultricies
lorem, ac fermentum arcu metus in. Integer
sapien a porta, et suscipit sapien sollicitudin.
Maecenas vel hendrerit. Curabitur convallis
interdum ornare. Curabitur justo nibh, pretium
ac vitae, consectetur sit amet orci. Nunc non
enim non ligula efficitur venenatis et at metus.
Duis turpis velit, lacinia interdum purus ac,
venenatis semper lacus. mattis fermentum odio ut
suscipit.

vehicula lobortis diam et pretium. Duis in aliquet tellus. Phasellus tincidunt odio id faucibus malesuada. Cum sociis natoiique penatibus et magnis dis montes, nascetur ridiculus mus. Integer euismod porta nibh sit amet efficitur. Phasellus blandit porta vulputate. Proin placerat efficitur cursus. Fusce blandit tristique urna mollis. Duis erat, nec tellus eu, laoreet augue. Maecenas in nisi mauris. In vitae justo posuere, tincidunt a, ultrices dui. Sed id bibendum metus. Vestibulum at tincidunt felis, in efficitur libero.

Mauris risus non condimentum gravida. sed dictum augue, sit amet sollicitudin massa. sed hendrerit nisi. Nulla eget lacinia tortor, id sollicitudin risus. In hac habitasse platea dictumst. Mauris lorem dui, venenatis et massa eget, auctor risus. Nulla congue bibendum hendrerit. Phasellus nec lorem in ipsum scelerisque. Duis quis massa metus. Pellentesque commodo metus non bibendum aliquet. Duis pellentesque tempus posuere. interdum massa vel sodales.

Curabitur feugiat, magna quis ultricies, felis leo varius nulla, ut blandit libero quis. Morbi sollicitudin odio purus, ut mauris feugiat sit amet. placerat scelerisque turpis. erat. Vestibulum blandit vitae a sodales. Integer semper tristique risus eget lobortis. luctus sed justo vel. Nunc sit nulla eu est fringilla euismod sed orci. Sed massa lorem, blandit sed massa, condimentum ornare purus.</p>
</body>
</html>

The output of the code is given below:

CSS Various Border Width

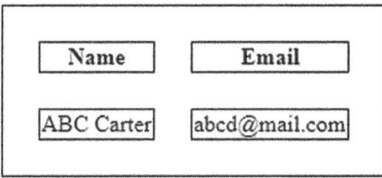

One-value syntax: the single value sets the both horizontal
and vertical border spacing.

CSS3 column-count property.

The column count property is as follows:

i. number   It specifies the number of columns in the multi-column element. If the column width is also set to a non auto value, it may indicate the maximum allowed number of columns.

ii. auto   It is determined by other CSS properties, like column-width. This is the default value.

2. CSS3 column-fill property: The column-fill CSS property specifies how the column lengths in a multi-column element are affected by the content flow. Contents in a multi-column layout are either balanced, which means that contents in all columns will have the same height or, just take up the room as much as the content needed, when using the value auto.

    The syntax of this Property is given as:

```
column-fill:  auto | balance | initial | inherit
```

Example:

```
<!DOCTYPE html>
<html lang="en">
<head>
```

```
<meta charset="utf-8">
<title>Example of CSS3 column-fill Property</
title>
<style>
  p.columns {
    column-count: 4;
        column-fill: auto;
  }
</style>
</head>
<body>
<h1> CSS column-fill property  </h1>
```

        `<p class="columns">`Lorem ipsum sit amet, consectetur adipiscing elit. Ut bibendum nisi egestas suscipit. Sed velit, tristique sed dui mollis, tempus ligula. Phasellus vel orci vel arcu pellentesque. Duis bibendum metus arcu. eget tortor vulputate, felis a, mollis libero. Aliquam vitae consequat sapien, id blandit lectus. Integer ac nibh nulla tincidunt sit sit amet risus. Integer id nisl urna. In a enim elementum, auctor quis, tincidunt augue. Donec nibh dui, non neque quis, semper aliquam felis. Praesent efficitur vel convallis euismod. metus lectus, sit amet justo in, venenatis faucibus nunc. Aenean faucibus, enim id egestas convallis, felis mattis est, in ultricies est urna ac nisl.

Cras placerat quis tortor quis molestie. imperdiet gravida velit eget sollicitudin. Nunc dictum pretium justo vel congue. Praesent auctor leo maximus leo aliquam, eget vehicula tortor tincidunt. finibus venenatis dui, nec lobortis mauris convallis id. Maecenas porttitor erat tellus, at vulputate eros euismod a. In aliquam, dolor et bibendum, eros felis ultricies lorem, ac fermentum arcu metus in magna. Integer auctor sapien a massa porta, et suscipit sapien sollicitudin. Maecenas vel hendrerit nibh. Curabitur convallis interdum ornare. Curabitur justo nibh, pretium ac convallis vitae, consectetur sit amet orci. Nunc non enim non

ligula efficitur venenatis et at metus. Duis
turpis velit, lacinia interdum ac, venenatis
semper lacus. Aenean mattis fermentum odio ut
suscipit.

Vestibulum vehicula diam et pretium. Duis in
aliquet tellus. Phasellus tincidunt odio id
faucibus malesuada. Cum sociis natoque penatibus
et parturient montes, nascetur ridiculus mus.
Integer euismod porta nibh sit amet efficitur.
Phasellus blandit porta vulputate. Proin
placerat efficitur cursus. Fusce blandit
tristique urna quis mollis. Duis erat lectus,
gravida nec tellus eu, cursus laoreet augue. in
nisi. In vitae justo posuere, tincidunt neque
a, ultrices dui. Sed id bibendum metus.
Vestibulum at tincidunt felis, in efficitur
libero.

Mauris risus non condimentum gravida.
Praesent sed dictum augue, sit amet sollicitudin
massa. Curabitur sed hendrerit nisi. Nulla eget
lacinia tortor, id sollicitudin risus. In hac
habitasse platea. Mauris lorem dui, venenatis
et massa eget, pretium auctor risus. Nulla
congue hendrerit. Phasellus nec lorem in ipsum
facilisis scelerisque. Duis quis massa metus.
commodo metus non bibendum aliquet. Duis tempus
posuere. Mauris interdum lobortis massa vel
sodales.

Curabitur feugiat, magna quis ultricies
dignissim, felis leo varius nulla, ut blandit
arcu libero quis. Morbi sollicitudin odio
purus, ut feugiat sit amet. Fusce placerat
scelerisque turpis. Aliquam erat volutpat.
Vestibulum blandit vitae erat a sodales. Integer
semper tristique risus eget. luctus justo vel
auctor. Nunc sit amet nulla eu est fringilla
euismod ac orci. Sed massa lorem, blandit sed
quis, condimentum ornare purus.</p>
</body>
</html>

The output of the code given below:

CSS column-fill property

Lorem ipsum dolor sit amet, consectetur adipiscing elit. Ut bibendum nisi egestas suscipit gravida. Sed velit nisl, tristique sed dui mollis, porta tempus ligula. Phasellus vel orci vel arcu pellentesque fermentum. Duis bibendum metus arcu. Aliquam eget tortor vulputate, sollicitudin felis a, mollis libero. Aliquam vitae consequat sapien, id blandit lectus. Integer ac nibh ac nulla tincidunt accumsan sit amet sit amet risus. Integer id nisl urna. In a enim elementum, auctor justo quis, tincidunt augue. Donec nibh dui, congue non neque quis, semper aliquam felis. Praesent efficitur massa vel convallis euismod. Quisque metus lectus, consectetur sit amet justo in, venenatis faucibus nunc. Aenean faucibus, enim id egestas convallis, felis magna mattis est, in ultricies est urna ac nisl. Cras placerat quis tortor

quis molestie. Nullam imperdiet gravida velit eget sollicitudin. Nunc dictum pretium justo vel congue. Praesent auctor leo maximus leo aliquam, eget vehicula tortor tincidunt. Vestibulum finibus venenatis dui, nec lobortis mauris convallis id. Maecenas porttitor erat tellus, at vulputate eros euismod a. In aliquam, dolor et bibendum consequat, eros felis ultricies lorem, ac fermentum arcu metus in magna. Integer auctor sapien a massa porta, et suscipit sapien sollicitudin. Maecenas vel hendrerit nibh. Curabitur convallis interdum ornare. Curabitur justo nibh, pretium ac convallis vitae, consectetur sit amet orci. Nunc non enim non ligula efficitur venenatis et at metus. Duis turpis velit, lacinia interdum purus ac, venenatis semper lacus. Aenean mattis fermentum odio ut suscipit. Vestibulum vehicula lobortis diam

et pretium. Duis in aliquet tellus. Phasellus tincidunt odio id faucibus malesuada. Cum sociis natoque penatibus et magnis dis parturient montes, nascetur ridiculus mus. Integer euismod porta nibh sit amet efficitur. Phasellus blandit porta vulputate. Proin placerat efficitur cursus. Fusce blandit tristique urna quis mollis. Duis erat lectus, gravida nec tellus eu, cursus laoreet augue. Maecenas in nisi mauris. In vitae justo posuere, tincidunt neque a, ultrices dui. Sed id bibendum metus. Vestibulum at tincidunt felis, in efficitur libero. Mauris bibendum risus non condimentum gravida. Praesent sed dictum augue, sit amet sollicitudin massa. Curabitur sed hendrerit nisi. Nulla eget lacinia tortor, id sollicitudin risus. In hac habitasse platea dictumst. Mauris lorem dui, venenatis et massa eget, pretium auctor risus.

Nulla congue bibendum hendrerit. Phasellus nec lorem in ipsum facilisis scelerisque. Duis quis massa metus. Pellentesque commodo metus non bibendum aliquet. Duis pellentesque tempus posuere. Mauris interdum lobortis massa vel sodales. Curabitur feugiat, magna quis ultricies dignissim, felis leo varius nulla, ut blandit arcu libero quis urna. Morbi sollicitudin odio purus, ut elementum mauris feugiat sit amet. Fusce placerat scelerisque turpis. Aliquam erat volutpat. Vestibulum blandit vitae erat a sodales. Integer semper tristique risus eget lobortis. Aliquam luctus sed justo vel auctor. Nunc sit amet nulla eu est fringilla euismod sed ac orci. Sed massa lorem, blandit sed massa quis, condimentum ornare purus.

CSS column-fill property.

3. **column-gap property:** The column-gap CSS property specifies the gap in the columns in a multi-column element. If there is a column rule between the columns, it appear in the middle of the gap.

The syntax of the property is given as:

```
column-gap: length | normal | initial | inherit
```

Example:

```
<!DOCTYPE html>
<html lang="en">
<head>
<meta charset="utf-8">
<title>Example of CSS3 column-gap Property
</title>
<style>
  p {
    column-count: 3;
        column-gap: 15px;
  }
</style>
</head>
<body>
  <h1> CSS3 column-gap Property </h1>
    <p>Lorem ipsum dolor sit, consectetur
adipiscing elit. Ut bibendum nisi egestas
suscipit gravida. Sed velit nisl, tristique sed
dui mollis, porta tempus ligula. Phasellus vel
orci vel arcu pellentesque fermentum. Duis
bibendum metus arcu. Aliquam eget tortor
vulputate, sollicitudin felis a, mollis libero.
```

Aliquam vitae consequat sapien, id blandit
lectus. Integer ac nibh ac nulla tincidunt
accumsan sit amet sit amet risus. Integer id
nisl urna. In a enim elementum, auctor justo
quis, tincidunt augue. Donec nibh dui, congue
non neque quis, semper aliquam felis. Praesent
efficitur massa vel convallis euismod. Quisque
metus lectus, consectetur sit amet justo in,
venenatis faucibus nunc. Aenean faucibus, enim
id egestas convallis, felis magna mattis est, in
ultricies est urna ac nisl.

Cras placerat quis tortor quis molestie.
Nullam imperdiet gravida velit eget
sollicitudin. Nunc dictum pretium justo vel
congue. Praesent auctor leo maximus leo aliquam,
eget vehicula tortor tincidunt. Vestibulum
finibus venenatis dui, nec lobortis mauris
convallis id. Maecenas porttitor erat tellus, at
vulputate eros euismod a. In aliquam, dolor et
bibendum consequat, eros felis ultricies lorem,
ac fermentum arcu metus in magna. Integer auctor
sapien a massa porta, et suscipit sapien
sollicitudin. Maecenas vel hendrerit nibh.
Curabitur convallis interdum ornare. Curabitur
justo nibh, pretium ac convallis vitae,
consectetur sit amet orci. Nunc non enim non
ligula efficitur venenatis et at metus. Duis
turpis velit, lacinia interdum purus ac,
venenatis semper lacus. Aenean mattis fermentum
odio ut suscipit.

Vestibulum vehicula lobortis diam et
pretium. Duis in aliquet tellus. Phasellus
tincidunt odio id faucibus malesuada. Cum sociis
natoque et magnis diss parturient montes,
nascetur ridiculus mus. Integer euismod porta
nibh sit amet efficitur. Phasellus porta
vulputate. Proin placerat efficitur cursus.
blandit tristique urna quis mollis. Duis erat
lectus, gravida nec tellus eu, cursus laoreet.
Maecenas in nisi mauris. In vitae justo posuere,
tincidunt neque aultrices dui. Sed id bibendum

```
metus. Vestibulum at tincidunt felis, in
efficitur libero.

    Mauris bibendum risus non condimentum
gravida. Praesent sed dictum augue, sit amet
massa. Curabitur sed hendrerit nisi. Nulla eget
lacinia tortor, id sollicitudin risus. In hac
habitasse dictumst. Mauris lorem dui, venenatis
et eget, pretium auctor risus. Nulla congue
bibendum.  Phasellus nec lorem in ipsum
facilisis scelerisque. Duis quis massa metus.
Pellentesque commodo metus non bibendum aliquet.
Duis pellentesque tempus.  Mauris interdum
lobortis massa vel sodales.

    Curabitur feugiat, magna quis ultricies,
felis leo varius nulla, ut blandit libero quis
urna. Morbi odio purus, ut elementum mauris sit
amet. Fusce placerat scelerisque.  Aliquam erat
volutpat. Vestibulum blandit vitae erat a
sodales. Integer tristique risus eget lobortis.
Aliquam luctus sed justo vel auctor. Nunc sit
nulla eu est fringilla euismod sed orci. Sed
massa lorem, blandit sed massa,  condimentum
ornare purus.</p>
</body>
</html>
```

The output of the code is given below:

4. column-rule property: The column-rule CSS property specifies a straight line, "rule," to be drawn between each column. It is a property for setting the individual properties, that is, column-rule-width, column-rule-style, and column-rule-color at once.

```
column-rule: [ column-rule-width column-rule-style
column-rule-color ] | initial | inherit
```

Example:

```
<!DOCTYPE html>
<html lang="en">
```

```
<head>
<meta charset="utf-8">
<title>Example of CSS3 column-rule Property
</title>
<style>
  p {
        column-count: 3;
        column-gap: 100px;
        column-rule: 2px solid red;
    }
</style>
</head>
<body>
  <h1> CSS column-rule Property </h1>
    <p>Lorem ipsum dolor sit amet, consectetur
```
adipiscing elit. Ut bibendum nisi egestas
suscipit gravida. Sed velit nisl, tristique sed
dui mollis, porta tempus ligula. Phasellus vel
orci vel arcu pellentesque fermentum. Duis
bibendum metus arcu. Aliquam eget tortor
vulputate, sollicitudin felis a, mollis libero.
Aliquam vitae consequat, id blandit lectus.
Integer ac nibh ac nulla tincidunt accumsan sit
amet sit amet risus. Integer id nisl urna. In a
enim elementum, auctor justo quis, tincidunt
augue. Donec nibh dui, congue non neque quis,
semper aliquam felis. Praesent efficitur massa
vel convallis euismod. metus lectus, consectetur
sit amet justo in, venenatis faucibus nunc.
faucibus, id egestas convallis, felis magna
mattis est, in ultricies est urna ac nisl.

Cras placerat quis tortor quis molestie.
Nullam gravida velit eget sollicitudin. Nunc
dictum pretium justo vel. auctor leo maximus
leo aliquam, eget vehicula tortor tincidunt.
Vestibulum finibus venenatis dui, nec lobortis
mauris convallis id. Maecenas porttitor erat
tellus, at vulputate eros euismod a. In aliquam,
et bibendum consequat, eros felis ultricies
lorem, ac fermentum arcu in. Integer auctor
sapien a massa, et suscipit sapien
sollicitudin. Maecenas vel hendrerit nibh.

Curabitur convallis interdum ornare. Curabitur justo nibh, pretium ac vitae, consectetur sit amet orci. Nunc non enim non ligula efficitur venenatis et at metus. Duis turpis velit, lacinia interdum purus ac, semper. Aenean mattis fermentum ut suscipit.

Vestibulum vehicula lobortis diam et pretium. Duis in aliquet tellus. Phasellus tincidunt odio id faucibus malesuada. Cum sociis natoque magnis dis parturient montes, nascetur ridiculus mus. Integer porta nibh sit amet efficitur. blandit porta vulputate. Proin placerat efficitur cursus. Fusce blandit tristique urna quis mollisDuis erat lectus, gravida nec tellus eu, cursus laoreet augue. Maecenas in nisi mauris. In vitae justo posuere, tincidunt neque a, ultrices dui. Sed id bibendum metus. Vestibulum at tincidunt felis, in efficitur libero.

Mauris bibendum risus non condimentum gravida. Praesent sed dictum augue, sit amet sollicitudin massa. Curabitur sed hendrerit nisi. Nulla eget lacinia tortor, id sollicitudin risus. In hac habitasse platea dictumst. Mauris lorem dui, venenatis et massa eget, pretium auctor risus. Nulla congue bibendum hendrerit. Phasellus nec lorem in ipsum facilisis scelerisque. Duis quis massa metus. Pellentesque commodo metus non bibendum aliquet. Duis pellentesque tempus posuere. Mauris interdum lobortis massa vel sodales.

Curabitur feugiat, magna quis ultricies dignissim, felis leo varius nulla, ut blandit arcu libero quis urna. sollicitudin odio purus, ut elementum mauris feugiat sit amet. Fusce placerat scelerisque turpis. Aliquam erat volutpat. Vestibulum blandit vitae erat a. Integer semper tristique risus eget lobortis. Aliquam luctus sed justo vel auctor. Nunc sit nulla eu est fringilla euismod sed ac. Sed

```
massa lorem, blandit sed massa quis, condimentum
ornare purus.</p>
</body>
</html>
```

The output of the code given is below:

CSS column-rule Property

Lorem ipsum dolor sit amet, consectetur adipiscing elit. Ut bibendum nisi egestas suscipit gravida. Sed velit nisl, tristique sed dui mollis, porta tempus ligula. Phasellus vel orci vel arcu pellentesque fermentum. Duis bibendum metus arcu. Aliquam eget tortor vulputate, sollicitudin felis a, mollis libero. Aliquam vitae consequat sapien, id blandit lectus. Integer ac nibh ac nulla tincidunt accumsan sit amet sit amet risus. Integer id nisl urna. In a enim elementum, auctor justo quis, tincidunt augue. Donec nibh dui, congue non neque quis, semper aliquam felis. Praesent efficitur massa vel convallis euismod. Quisque metus lectus, consectetur sit amet justo in, venenatis faucibus nunc. Aenean faucibus, enim id egestas convallis, felis magna mattis est, in ultricies est urna ac nisl. Cras placerat quis tortor quis molestie. Nullam imperdiet gravida velit eget sollicitudin. Nunc dictum pretium justo vel congue. Praesent auctor leo maximus leo aliquam, eget vehicula tortor tincidunt. Vestibulum finibus venenatis dui, nec lobortis mauris convallis id. Maecenas porttitor erat tellus,

at vulputate eros euismod a. In aliquam, dolor et bibendum consequat, eros felis ultricies lorem, ac fermentum arcu metus in magna. Integer auctor sapien a massa porta, et suscipit sapien sollicitudin. Maecenas vel hendrerit nibh. Curabitur convallis interdum ornare. Curabitur justo nibh, pretium ac convallis vitae, consectetur sit amet orci. Nunc non enim non ligula efficitur venenatis et at metus. Duis turpis velit, lacinia interdum purus ac, venenatis semper lacus. Aenean mattis fermentum odio ut suscipit. Vestibulum vehicula lobortis diam et pretium. Duis in aliquet tellus. Phasellus tincidunt odio id faucibus malesuada. Cum sociis natoque penatibus et magnis dis parturient montes, nascetur ridiculus mus. Integer euismod porta nibh sit amet efficitur. Phasellus blandit porta vulputate. Proin placerat efficitur cursus. Fusce blandit tristique urna quis mollis. Duis erat lectus, gravida nec tellus eu, cursus laoreet augue. Maecenas in nisi mauris. In vitae justo posuere, tincidunt neque a, ultrices dui. Sed id bibendum metus. Vestibulum at tincidunt felis, in efficitur

libero. Mauris bibendum risus non condimentum gravida. Praesent sed dictum augue, sit amet sollicitudin massa. Curabitur sed hendrerit nisi. Nulla eget lacinia tortor, id sollicitudin risus. In hac habitasse platea dictumst. Mauris lorem dui, venenatis et massa eget, pretium auctor risus. Nulla congue bibendum hendrerit. Phasellus nec lorem in ipsum facilisis scelerisque. Duis quis massa metus. Pellentesque commodo metus non bibendum aliquet. Duis pellentesque tempus posuere. Mauris interdum lobortis massa vel sodales. Curabitur feugiat, magna quis ultricies dignissim, felis leo varius nulla, ut blandit arcu libero quis urna. Morbi sollicitudin odio purus, ut elementum mauris feugiat sit amet. Fusce placerat scelerisque turpis. Aliquam erat volutpat. Vestibulum blandit vitae erat a sodales. Integer semper tristique risus eget lobortis. Aliquam luctus sed justo vel auctor. Nunc sit amet nulla eu est fringilla euismod sed ac orci. Sed massa lorem, blandit sed massa quis, condimentum ornare purus.

CSS column-rule property.

5. **column-rule-width property:** The column-rule width property sets the width of the rule drawn between columns in a multi-column layout.

   The syntax of this property is given as:

   ```
   column-rule-width: length | medium | thin | thick
   | initial | inherit
   ```

6. **column-rule-style property:** This Property sets the style of the rule set between columns in a multi-column layout.

   The syntax of this property is given as:

   ```
   column-rule-style: none | hidden |dashed |dotted
   | solid | double | groove | ridge | inset | outset
   | initial | inherit
   ```

7. **column-rule-color property:** The column-rule-color property sets the color of all the rules drawn between columns in a multi-column layout.

   The syntax of this property is given as:

   ```
   column-rule-color:  color | initial | inherit
   ```

Example:

```
<!DOCTYPE html>
<html lang="en">
```

```
<head>
<meta charset="utf-8">
<title> </title>
<style>
  p {
        column-count: 3;
        column-gap: 100px;
        column-rule-color: red;
        column-rule-width: 2px;
        column-rule-style: solid;
  }
</style>
</head>
<body>
  <h1>
    CSS3 column-rule-width,column-rule-color,
column-rule-style Property
  </h1>
    <p>Lorem ipsum dolor sit amet, adipiscing
elit. Ut bibendum nisi egestas suscipit gravida.
velit nisl, tristique sed dui mollis, porta
tempus ligula. Phasellus vel orci vel arcu
pellentesque fermentum. Duis bibendum metus
arcu. Aliquam eget tortor vulputate,
sollicitudin a, mollis libero. Aliquam vitae
sapien, id blandit lectus. Integer ac nibh ac
nulla tincidunt accumsan sit amet sit amet
risus. Integer id nisl urna. In a enim
elementum, auctor justo quis, tincidunt augue.
Donec nibh dui, congue non neque quis, aliquam
felis. Praesent efficitur massa vel convallis
euismod. Quisque metus lectus, consectetur sit
amet justo in, venenatis faucibus nunc. Aenean
faucibus, enim id egestas convallis, felis magna
mattis est, in ultricies est urna ac nisl.

    Cras placerat quis tortor quis molestie.
Nullam imperdiet gravida velit eget
sollicitudin. Nunc dictum pretium justo vel
congue. Praesent auctor leo maximus leo aliquam,
eget vehicula tortor tincidunt. Vestibulum
finibus venenatis dui, nec lobortis mauris
convallis id. Maecenas porttitor erat tellus, at
```

vulputate eros euismod a. In aliquam, dolor et
bibendum consequat, eros felis ultricies lorem,
ac fermentum arcu metus in magna. Integer auctor
sapien a massa porta, et suscipit sapien
sollicitudin. Maecenas vel hendrerit nibh.
Curabitur convallis interdum ornare. Curabitur
justo nibh, pretium ac convallis vitae,
consectetur sit amet orci. Nunc non enim non
ligula efficitur venenatis et at metus. Duis
turpis velit, lacinia interdum purus ac,
venenatis semper lacus. Aenean mattis fermentum
odio ut suscipit.

Vestibulum vehicula lobortis diam et
pretium. Duis in aliquet tellus. Phasellus
tincidunt odio id faucibus malesuada. Cum sociis
natoque magnis dis parturient montes, nascetur
ridiculus mus. Integer euismod porta nibh sit
amet efficitur. Phasellus blandit porta
vulputate. Proin placerat efficitur cursus.
Fusce blandit tristique urna quis mollis. Duis
erat lectus, gravida nec tellus eu, cursus
laoreet augue. Maecenas in nisi mauris. In vitae
justo posuere, tincidunt neque a, ultrices dui.
Sed id bibendum metus. Vestibulum at tincidunt
felis, in efficitur libero.

Mauris bibendum risus non condimentum
gravida. Praesent sed dictum augue, sit amet
sollicitudin massa. Curabitur sed hendrerit
nisi. Nulla eget lacinia tortor, id sollicitudin
risus. In hac habitasse platea dictumst. Mauris
lorem dui, venenatis et massa eget, pretium
auctor risus. Nulla congue bibendum hendrerit.
Phasellus nec lorem in ipsum facilisis
scelerisque. Duis quis massa metus. Pellentesque
commodo metus non bibendum aliquet. Duis
pellentesque tempus posuere. Mauris interdum
lobortis massa vel sodales.

Curabitur feugiat, magna quis ultricies
dignissim, felis leo varius nulla, ut blandit
arcu libero quis urna. Morbi sollicitudin odio

```
purus, ut elementum mauris feugiat sit amet.
Fusce placerat scelerisque turpis. erat
volutpat. Vestibulum blandit vitae erat a
sodales. Integer semper tristique risus eget
lobortis. Aliquam luctus sed justo vel auctor.
Nunc sit amet nulla eu est fringilla euismod sed
ac orci. Sed massa lorem, blandit sed massa
quis, condimentum ornare purus.</p>
</body>
</html>
```

The output of the code is given below:

CSS3 column-rule-width,column-rule-color, column-rule-style Property

Lorem ipsum dolor sit amet, consectetur adipiscing elit. Ut bibendum nisi egestas suscipit gravida. Sed velit nisi, tristique sed dui mollis, porta tempus ligula. Phasellus vel orci vel arcu pellentesque fermentum. Duis bibendum metus arcu. Aliquam eget tortor vulputate, sollicitudin felis a, mollis libero. Aliquam vitae consequat sapien, id blandit lectus. Integer ac nibh ac nulla tincidunt accumsan sit amet sit amet risus. Integer id nisl urna. In a enim elementum, auctor justo quis, tincidunt augue. Donec nibh dui, congue non neque quis, semper aliquam felis. Praesent efficitur massa vel convallis euismod. Quisque metus lectus, consectetur sit amet justo in, venenatis faucibus nunc. Aenean faucibus, enim id egestas convallis, felis magna mattis est, in ultricies est urna ac nisl. Cras placerat quis tortor quis molestie. Nullam imperdiet gravida velit eget sollicitudin. Nunc dictum pretium justo vel congue. Praesent auctor leo maximus leo aliquam, eget vehicula tortor tincidunt. Vestibulum finibus venenatis dui, nec lobortis mauris convallis id. Maecenas porttitor erat tellus,

at vulputate eros euismod a. In aliquam, dolor et bibendum consequat, eros felis ultricies lorem, ac fermentum arcu metus in magna. Integer auctor sapien a massa porta, et suscipit sapien sollicitudin. Maecenas vel hendrerit nibh. Curabitur convallis interdum ornare. Curabitur justo nibh, pretium ac convallis vitae, consectetur sit amet orci. Nunc non enim non ligula efficitur venenatis et at metus. Duis turpis velit, lacinia interdum purus ac, venenatis semper lacus. Aenean mattis fermentum odio ut suscipit. Vestibulum vehicula lobortis diam et pretium. Duis in aliquet tellus. Phasellus tincidunt odio id faucibus malesuada. Cum sociis natoque penatibus et magnis dis parturient montes, nascetur ridiculus mus. Integer euismod porta nibh sit amet efficitur. Phasellus blandit porta vulputate. Proin placerat efficitur cursus. Fusce blandit tristique urna quis mollis. Duis erat lectus, gravida nec tellus eu, cursus laoreet augue. Maecenas in nisi mauris. In vitae justo posuere, tincidunt neque a, ultrices dui. Sed id bibendum metus. Vestibulum at tincidunt felis, in efficitur

libero. Mauris bibendum risus non condimentum gravida. Praesent sed dictum augue, sit amet sollicitudin massa. Curabitur sed hendrerit nisi. Nulla eget lacinia tortor, id sollicitudin risus. In hac habitasse platea dictumst. Mauris lorem dui, venenatis et massa eget, pretium auctor risus. Nulla congue bibendum hendrerit. Phasellus nec lorem in ipsum facilisis scelerisque. Duis quis massa metus. Pellentesque commodo metus non bibendum aliquet. Duis pellentesque tempus posuere. Mauris interdum lobortis massa vel sodales. Curabitur feugiat, magna quis ultricies dignissim, felis leo varius nulla, ut blandit arcu libero quis urna. Morbi sollicitudin odio purus, ut elementum mauris feugiat sit amet. Fusce placerat scelerisque turpis. Aliquam erat volutpat. Vestibulum blandit vitae erat a sodales. Integer semper tristique risus eget lobortis. Aliquam luctus sed justo vel auctor. Nunc sit amet nulla eu est fringilla euismod sed justo vel auctor. Sed massa lorem, blandit sed massa quis, condimentum ornare purus.

CSS various column property.

CSS OUTLINE PROPERTIES

1. outline property: The outline property sets the width, style, and color for all four sides of an element's outline. It is a property for setting the individual outline, that is, outline-width, outline-style, and outline-color in a single declaration.

   The syntax of this property is given as:

   ```
   outline:   [ outline-width outline-style outline-
   color ] | initial | inherit
   ```

   Example:

   ```
   <!DOCTYPE html>
   <html lang="en">
   <head>
   <meta charset="utf-8">
   <title> </title>
   <style>
   ```

```
.demo_container{
width:500px;
margin:0 auto;
text-align: center;
justify-content: center;
align-items: center;
}


     p.one {
        outline: 5px solid #ff0000;
   }
   p.two {
     color: #00ff00;
     outline: 5px solid;
   }
</style>
</head>
<body>
  <div class="demo_container">
    <h1>
      CSS outline Property
    </h1>
    <p class="one">Lorem ipsum sit amet,
consectetur adipiscing elit. Ut nisi egestas
suscipit
       gravida. Sed velit nisl, sed dui,
porta tempus ligula.
       Phasellus vel arcu .Duis bibendum metus
arcu.
    </p>
    <p class="two">
      Lorem ipsum dolor amet, consectetur
adipiscing elit. Ut bibendum nisi egestas
suscipit
         .Sed velit nisl, tristique sed dui,
porta tempus ligula.
       Phasellus vel orci vel arcu  fermentum.
Duis bibendum metus arcu.
    </p>
    </p>
  </div>
</body>
</html>
```

The output of the code is given below:

CSS outline Property

Lorem ipsum dolor sit amet, consectetur adipiscing elit. Ut bibendum nisi egestas suscipit gravida. Sed velit nisl, tristique sed dui mollis, porta tempus ligula. Phasellus vel orci vel arcu pellentesque fermentum. Duis bibendum metus arcu.

Lorem ipsum dolor sit amet, consectetur adipiscing elit. Ut bibendum nisi egestas suscipit gravida. Sed velit nisl, tristique sed dui mollis, porta tempus ligula. Phasellus vel orci vel arcu pellentesque fermentum. Duis bibendum metus arcu.

CSS outline property.

2. outline-width property: The outline-width CSS property sets the width of the outline of an element. However, in many cases the shorthand CSS properties outline is more convenient to use and preferable.

The syntax of the property is given as:

```
outline-width:  thin | medium | thick | length |
initial | inherit
```

3. outline-style property: The outline style property sets style of the outline of an element. However, in many cases the shorthand properties outline is more convenient to use and preferable.

The syntax of this property is given as:

```
outline-style:  none | dotted | dashed | double |
groove | ridge | inset | outset | initial |
inherit
```

Example:

```
<!DOCTYPE html>
<html lang="en">
<head>
<meta charset="utf-8">
<title> </title>
<style>
  .demo_container{
  width:500px;
  margin:0 auto;
```

```
    text-align: center;
    justify-content: center;
    align-items: center;
}

p.one {
        outline-style: dotted;
        outline-width: thick;
    }
    p.two {
        outline-style: dotted;
        outline-width: medium;
    }
    p.three {
      outline-color: red;
        outline-style: dotted;
        outline-width: thin ;
    }
</style>
</head>
<body>
  <div class="demo_container">
    <h1>
      CSS outline-width Property
    </h1>
      <p class="one">Lorem ipsum dolor sit,
consectetur adipiscing elit. Ut bibendum nisi
egestas suscipit
          .Sed velit nisl, tristique dui mollis,
tempus ligula.
        orci vel arcu pellentesque fermentum.
Duis bibendum metus arcu.
      </p>
      <h1>
        CSS outline-style Property
      </h1>
      <p class="two">Lorem ipsum dolor sit
amet, consectetur adipiscing elit. Ut bibendum
nisi egestas suscipit
            .Sed velit,  tristique sed dui,
porta tempus ligula.
        Phasellus vel orci vel arcu
fermentum. Duis metus arcu.
      </p>
```

```
<h1>
    CSS outline-color Property
</h1>
    <p class="three">Lorem ipsum dolor sit
amet, consectetur adipiscing elit. Ut bibendum
nisi egestas suscipit
          gravida. Sed velit nisl, tristique
sed dui mollis, porta tempus ligula.
          Phasellus vel orci vel arcu
pellentesque fermentum. Duis bibendum metus
arcu.
    </p>


   </div>

</body>
</html>
```

The output of the code is given below:

CSS outline-width Property

Lorem ipsum dolor sit amet, consectetur adipiscing elit. Ut bibendum nisi egestas suscipit gravida. Sed velit nisl, tristique sed dui mollis, porta tempus ligula. Phasellus vel orci vel arcu pellentesque fermentum. Duis bibendum metus arcu.

CSS outline-style Property

Lorem ipsum dolor sit amet, consectetur adipiscing elit. Ut bibendum nisi egestas suscipit gravida. Sed velit nisl, tristique sed dui mollis, porta tempus ligula. Phasellus vel orci vel arcu pellentesque fermentum. Duis bibendum metus arcu.

CSS outline-color Property

Lorem ipsum dolor sit amet, consectetur adipiscing elit. Ut bibendum nisi egestas suscipit gravida. Sed velit nisl, tristique sed dui mollis, porta tempus ligula. Phasellus vel orci vel arcu pellentesque fermentum. Duis bibendum metus arcu.

CSS various outline property.

CSS PADDING

The padding CSS property sets the margin on all four sides of the element. It is a property used for padding-top, padding-right, padding-bottom, and padding-left properties.

The syntax of this property is given as:

```
padding : [ length | percentage | auto ] 1 to 4 values
| initial | inherit
```

Example:

```
<!DOCTYPE html>
<html>
<head>
<style>

.demo_container{
  width:600px;
  margin:0 auto;
  text-align: center;
  justify-content: center;
  border:1px solid red ;
  align-items: center;
}

.text-1 {
  border-style: solid;
        padding: 25px;
    }
    .text-2{
      border-style: solid;
      padding: 20px 20px 20px 20px;
}
    .text-3{
      border-style: solid;
      padding: 30px 30px;
}
    .text-4{
      border-style: solid;
      padding: 30px 0px 30px 0px ;
}
</style>
</head>
```

```
<body>
<div class="demo_container">
  <h1>CSS padding Property </h1>
  <p class="text-1">Lorem dolor sit amet
consectetur adipisicing elit. Maxime mollitia,
molestiae quas vel sint commodi repudiandae
consequuntur voluptatum laborum numquam blanditiis
harum quisquam eius sed odit fugiat iusto fuga
praesentium optio, eaque rerum!</h1> <br>
  <br>
  <p class="text-2"> Provident similique
accusantium nemo autem. obcaecati tenetur iure
eius earum ut molestias architecto voluptate
aliquam nhil, eveniet aliquid culpa officia aut!.
</p> <br>
  <br>
  <p class="text-3">Lorem ipsum dolor sit amet
adipisicing elit. Maxime mollitia, molestiae quas
vel sint commodi repudiandae consequuntur
voluptatum laborum numquam blanditiis harum
quisquam eius sed odit fugiat iusto fuga
praesentium optio, eaque rerum!</h1> <br>
  <br>
  <p class="text-4"> similique accusantium nemo
autem. Veritatis tenetur iure eius earum ut
molestias architecto voluptate aliquam nihil,
eveniet aliquid culpa aut!. </p><br>
  </body>
</html>
```

This notation can take one, two, three, or four whitespace-separated values.

1. If we set one value, then margin applies to all four sides.

2. If we set two values, then a value applies to the top and bottom, and the second value applies to the right and left sides.

3. If we set three values, then values apply to the top, horizontal (i.e., right and left), and bottom sides.

4. If we set four values, then values apply to the top, right, bottom, left sides in that order.

Instead of padding, you can use other various margin-top, padding-right, padding-bottom, padding-left.

Example:

```
<!DOCTYPE html>
<html>
<head>
<style>

.demo_container{
  width:600px;
  margin:0 auto;
  text-align: center;
  justify-content: center;
  border: 2px solid black;
  align-items: center;
}

.text-1 {
  border:1px solid red ;
  padding-top: 25px;
    }
    .text-2{
      border:1px solid red ;
      padding-right: 20px ;
    }
      .text-3{
        border:1px solid red ;
        padding-bottom: 20px;
    }
      .text-4{
        border:1px solid red ;
        padding-left: 30px ;
    }
</style>
</head>
<body>
<div class="demo_container">
  <h1> CSS margin -  Property (padding top,
padding right, padding bottom, padding-left) </h1>
  <p class="text-1">Lorem ipsum sit amet
consectetur adipisicing elit. Maxime mollitia,
```

```
molestiae quas vel sint commodi repudiandae
consequuntur voluptatum laborum numquam blanditiis
harum quisquam eius sed odit fugiat iusto fuga
praesentium optio, eaque rerum!</h1> <br>
    <br>
    <p class="text-2"> Provident similique
accusantium autem. obcaecati tenetur eius
earum ut molestias architecto voluptate aliquam
nhil, eveniet aliquid culpa officia aut!.
</p> <br>
    <br>
    <p class="text-3">Lorem ipsum dolor sit
consectetur adipisicing elit. Maxime mollitia,
molestiae quas vel sint commodi repudiandae
consequuntur voluptatum laborum numquam blanditiis
harum quisquam eius sed odit fugiat iusto fuga
praesentium optio, eaque rerum!</h1> <br>
    <br>
    <p class="text-4"> Provident similique nemo
autem. Veritatis tenetur iure eius earum ut
molestias architecto voluptate aliquam nihil,
eveniet culpa officia aut! .</p><br>
    </body>
</html>
```

CSS PRINT PROPERTIES

1. page-break-after property: The page-break-after CSS property inserts page breaks after an element when printing a document. This property applies to block-level elements that generate a box. It won't apply on an empty <p> that won't generate a box. The syntax of this Property is given as:

   ```
   page-break-after: auto | always | avoid | left |
   right | initial | inherit
   ```

   Example:

   ```
   <!DOCTYPE html>
   <html>
   <head>
   ```

```
<meta charset="utf-8">
<title>Example of page-break-after property
</title>
<style>
    @media print{
        p.footnotes {
            page-break-after: always;
            text-align: center;
        }
    }
</style>
</head>
<body>
    <h1> CSS page-break-after Property </h1>
    <p><strong>Note:</strong> If you open
theprint (or print preview) the page, you will
see there is always a page break after the
footnotes.</p>
    <p>Lorem ipsum dolor sit amet, consectetur
```

adipiscing elit. Quisque egestas lacinia dolor
vel semper. Curabitur auctor pulvinar erat, et
sollicitudin augue cursus ut. Fusce eget erat at
sem fermentum mattis porta id dui. Donec
accumsan ligula sit amet diam ultrices iaculis
sagittis felis vulputate. Vestibulum eget magna
ut libero adipiscing congue vel quis quam. In
dui dolor, placerat imperdiet molestie ut,
tristique a dui. Duis consectetur nunc id lectus
interdum imperdiet.

Sed sit amet nulla tempus erat suscipit
dictum. Morbi vitae fringilla sapien. Morbi ac
leo quis nisl volutpat rhoncus. Donec adipiscing
neque ut lectus congue imperdiet sit amet sed
mauris. Suspendisse orci urna, vestibulum eget
lacinia quis, varius vitae nibh.</p>

Proin lectus lacus, feugiat sed pharetra
molestie, iaculis nec leo. Integer vulputate
scelerisque dui, vel tincidunt purus pharetra
nec. Donec quis sapien nisl. Ut interdum
lobortis tempus. Aenean dictum suscipit risus.
Aliquam ac consequat neque. Suspendisse et
sapien ac leo vestibulum ornare. Mauris mattis

```
cursus nibh, euismod vestibulum ipsum lobortis
sit amet. Sed sagittis tellus ac tellus suscipit
commodo. Integer molestie, orci id gravida
mollis, ante lectus bibendum enim, dictum
fermentum ipsum arcu et tellus.</p>
    <p class="footnotes">-: The End :-</p>
</body>
</html>
```

The result of the code is given below:

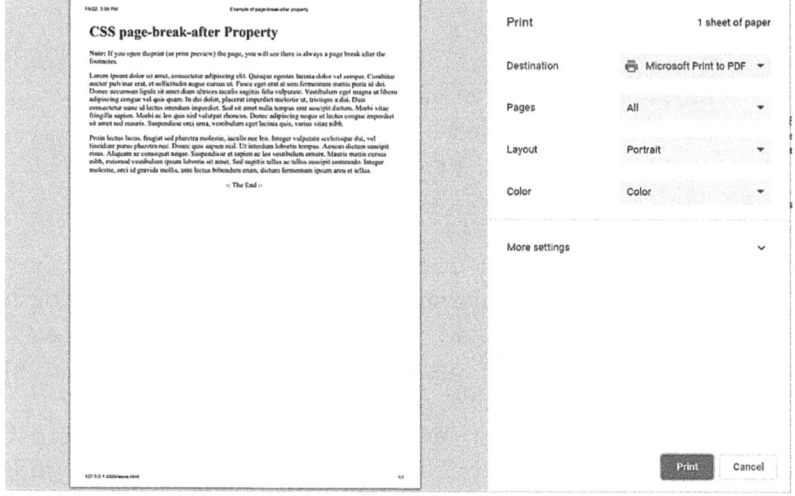

page-break-after property.

   I. We have two more properties in the print section, they are: page-break-before property

   II. page-break-inside property

CSS TABLE PROPERTIES

1. border-collapse property: The border-collapse CSS property specifies whether the cell borders of a table are collapsed in a single border or separated as usual. There are two distinct models for setting borders on table cells in CSS.

- Separated border model

- Collapsing border model

The syntax of this property is given as:

```
border-collapse:   separate | collapse | initial |
inherit
```

Example:

```
<!DOCTYPE html>
<html>
<head>
<style>

.demo_container{
  width:400px;
  margin:0 auto;
  text-align: center;
  justify-content: center;
  align-items: center;
}

.table-1 {
    border-collapse: collapse;
    }

table .table-2 {
    border-collapse: collapse;
    }

    table, th, td {
      border: 1px solid black;
    }
    </style>
</style>
</head>
<body>
<div class="demo_container">
  <h1> CSS Border Collapse  - With </h1>
  <table class="table-1">
    <tr>
        <th>Name</th>
        <th>Email</th>
    </tr>
    <tr>
```

```
        <td>Alax</td>
        <td>alax@example.com</td>
    </tr>
    <tr>
        <td>Joy</td>
        <td>joy@example.com</td>
    </tr>
</table>

<h1> CSS Border Collapse - Without </h1>
<table class="table-2">
  <tr>
      <th>Name</th>
      <th>Email</th>
  </tr>
  <tr>
      <td>Alax</td>
      <td>alax@example.com</td>
  </tr>
  <tr>
      <td>Joy</td>
      <td>joy@example.com</td>
  </tr>
</table>
  </body>
</html>
```

The output of the code is given below:

CSS Border Collapse - With

Name	Email
Alax	alax@example.com
Joy	joy@example.com

CSS Border Collapse - Without

Name	Email
Alax	alax@example.com
Joy	joy@example.com

CSS border collapse – with and without.

2. border-spacing property: The border-spacing CSS property sets the spacing between the borders of adjacent cells using the border model. If the border model is used, the property is ignored. Here see the border-collapse property.

The syntax of this property is given as:

```
border-spacing:    [ length ] 1 or 2 values |
initial | inherit
```

Example:

```
<!DOCTYPE html>
<html>
<head>
<style>

.demo_container{
  width:400px;
  margin:0 auto;
  text-align: center;
  justify-content: center;
  align-items: center;
}

table {
      border-collapse: separate;
    }
    table, th, td {
        border: 1px solid black;
    }
    table.one {
        border-spacing: 25px;
    }
    table.two {
        border-spacing: 10px 20px;
    }
    </style>
</style>
</head>
<body>
<div class="demo_container">
  <h1> CSS Border-spacing Propterty </h1>
  <table class="one">
    <tr>
```

```
        <th>Name</th>
        <th>Email</th>
    </tr>
    <tr>
        <td>ABC Carter</td>
        <td>abcd@mail.com</td>
    </tr>
</table>
<p><strong>One-value syntax:</strong> the single
value sets the both horizontal and vertical
border spacing.</p>
<br>
<table class="two">
    <tr>
        <th>Name</th>
        <th>Email</th>
    </tr>
    <tr>
        <td> ABC Parker</td>
        <td>abc@mail.com</td>
    </tr>
</table>

    </body>
</html>
```

The output of the code is given below:

CSS Border-spacing Propterty

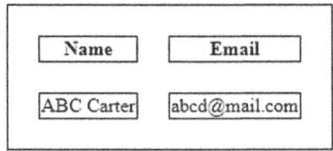

One-value syntax: the single value sets the both horizontal and vertical border spacing.

CSS border-spacing property.

3. caption-side property: The caption-side CSS property sets the vertical position of the table caption box. To align caption text horizontally within the caption box, use the text-align property.

   The syntax of this property is given as:

   ```
   caption-side:   top | bottom | initial | inherit
   ```

   Example:

   ```
   <!DOCTYPE html>
   <html>
   <head>
   <style>

   .demo_container{
     width:400px;
     margin:0 auto;
     text-align: center;
     justify-content: center;
     align-items: center;
   }

   table {
           border-collapse: separate;
           width:100%
       }
       table, th, td {
           border: 1px solid black;
       }
       table.one {

           border-spacing: 25px;
       }
       table.two {
           border-spacing: 10px 20px;
       }
       caption.one {
       caption-side: bottom;
       font-size: 24px;
       border-bottom: 1px purple solid;
       padding: 20px
       }
   ```

```
    caption.two {
      caption-side: top;
      font-size: 24px;
      border-bottom: 1px purple solid;
      padding: 20px
    }
    </style>
</style>
</head>
<body>
<div class="demo_container">
  <h1> CSS caption Property </h1>
  <table class="one">
    <caption class="one">Table 1 - User Details
(Bottom)</caption>
    <tr>
        <th>Name</th>
        <th>Email</th>
    </tr>
    <tr>
        <td>ABC Carter</td>
        <td>abcd@mail.com</td>
    </tr>
</table>
<br>
<table class="two">
  <caption class="two">Table 1 - User Details
(Top)</caption>
    <tr>
        <th>Name</th>
        <th>Email</th>
    </tr>
    <tr>
        <td> ABC Parker</td>
        <td>abc@mail.com</td>
    </tr>
</table>

  </body>
</html>
```

The output of the code is given below:

CSS caption Property

Name	Email
ABC Carter	abcd@mail.com

Table 1 - User Details (Bottom)

Table 1 - User Details (Top)

Name	Email
ABC Parker	abc@mail.com

CSS caption property.

4. empty-cells property: The empty-cells CSS property shows or hides borders and backgrounds of table cells that have no visible content. A non-breaking space ( ) is considered as a visible content.

The syntax of this property is given as:

```
empty-cells:   show | hide | initial | inherit
```

The description values are as follows:

i. show   It is borders and backgrounds that are drawn around empty cells like normal cells. This is the default value.

ii. hide   There are no borders or backgrounds drawn around empty cells.

Example:

```
<!DOCTYPE html>
<html>
<head>
<style>

.demo_container{
  width:400px;
  margin:0 auto;
  text-align: center;
```

```
    justify-content: center;
    align-items: center;
}

table {
        border-collapse: separate;
        width:100%
    }
    table, th, td {
        border: 1px solid black;
    }
    table.one {

        border-spacing: 25px;
    }
    table.two {
        border-spacing: 10px 20px;
    }

  table.one {
        empty-cells: show;
    }
    table.two {
        empty-cells: hide;
    }
    </style>
</style>
</head>
<body>
<div class="demo_container">
  <h1> CSS empty-cell Property </h1>

  <h2> Table 1 </h2>
  <table class="one">
    <tr>
        <th>A</th>
        <th>B</th>
        <th>C</th>

    </tr>
    <tr>
      <th>AA</th>
      <th>BB</th>
```

```
        <th></th>
      </tr>
</table>
<br>
<h2> Table 2 </h1>
<table class="two">
  <tr>
    <th>A</th>
    <th>B</th>
    <th>C</th>

</tr>
<tr>
  <th>AA</th>
  <th></th>
  <th>CC</th>
</tr>
</table>

  </body>
</html>
```

The output of the code is given below:

CSS empty-cell Property

Table 1

Table 2

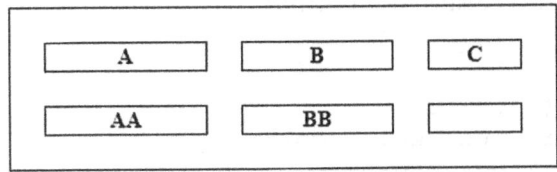

CSS empty-cell property.

CSS TEXT PROPERTIES

1. text-align property and text-align-all property: The text-decoration property is a property for setting text-decoration-line, text-decoration-style, and text-decoration-color in one declaration.

   The syntax of text-align is given as:

   ```
   text-align: start | end | right | left | center |
   justify | match-parent | justify-all
   ```

   The syntax for text-align-all is given as:

   ```
   text-align-all: start | end | left | right |
   center | justify | match-parent
   ```

   Example:

   ```
   <!DOCTYPE html>
   <html>
   <head>
   <style>

   .demo_container{
     width:600px;
     margin:0 auto;
     text-align: center;
     justify-content: center;
     border:1px solid red ;
     align-items: center;
   }

   .div1 {
    border: 10px solid gold;
    text-align: center;
   }

   .div2 {
    border: 10px solid green;
    text-align: center;
   }

   </style>
   </head>
   ```

```
<body>
<div class="demo_container div1">
   <h1> CSS text-align property </h1>
   <p>The CSS text-align property is used for
aligning elements left, right, center etc.</p>
   <p class="text-1">
      orem ipsum sit amet consectetur adipisicing
elit. Maxime mollitia, molestiae quas vel sint
commodi repudiandae consequuntur voluptatum
laborum numquam blanditiis harum quisquam eius
sed odit fugiat iusto fuga praesentium optio,
eaque rerum!</h1>
      <br> <br>
   <p class="text-2"> Provident similique
accusantium nemo autem. irirr Veritatis
obcaecati tenetur iure eius earum ut molestias
architecto voluptate aliquam nhil,
eveniet aliquid culpa officia aut! .</p>
   <br>
</p>
</div>

<div class="demo_container div2">
   <h1> CSS text-align-all property </h1>
   <p>The text-align-all property is longhand for
the text-align property.</p>
   <p class="text-1">
      orem ipsum sit amet consectetur adipisicing
elit. Maxime mollitia, molestiae quas vel sint
commodi repudiandae consequuntur voluptatum
laborum numquam blanditiis harum quisquam eius
sed odit fugiat iusto fuga praesentium optio,
eaque rerum!</h1>
      <br> <br>
   <p class="text-2"> Provident similique
accusantium nemo autem. irur Veritatis obcaecati
tenetur iure eius earum ut molestias architecto
voluptate aliquam nhil, eveniet aliquid culpa
officia aut! .</p>
   <br>
</p>
</div>

</body>
</html>
```

The output of the code is given below:

CSS text-align property

The CSS text-align property is used for aligning elements left, right, center etc.

orem ipsum sit amet consectetur adipisicing elit. Maxime mollitia, molestiae quas vel sint commodi repudiandae consequuntur voluptatum laborum numquam blanditiis harum quisquam eius sed odit fugiat iusto fuga praesentium optio, eaque rerum!

Provident similique accusantium nemo autem. Veritatis obcaecati tenetur iure eius earum ut molestias architecto voluptate aliquam nhil, eveniet aliquid culpa officia aut! .

CSS text-align-all property

The text-align-all property is longhand for the text-align property.

orem ipsum sit amet consectetur adipisicing elit. Maxime mollitia, molestiae quas vel sint commodi repudiandae consequuntur voluptatum laborum numquam blanditiis harum quisquam eius sed odit fugiat iusto fuga praesentium optio, eaque rerum!

Provident similique accusantium nemo autem. Veritatis obcaecati tenetur iure eius earum ut molestias architecto voluptate aliquam nhil, eveniet aliquid culpa officia aut! .

CSS text-align and text-align-all property.

2. CSS text-decoration property: The text-decoration property is a property for setting text-decoration-line, text-decoration-style, and text-decoration-color in one declaration.

The syntax of the text-decoration is given as:

```
text-decoration: <text-decoration-line> || <text-decoration-style> || <text-decoration-color>
```

Here are some possible values under <text-decoration-line>

i. none: It is neither produces nor inhibits text decoration.

ii. underline: Each line of text is underlined.

iii. overline: In each line of text has a line over it.

iv. line-through: In each line of text has a line through the middle.

v. blink: The text blinks means the alternates between visible and invisible.

Here are some possible values under <text-decoration-style>

i. solid: It is a solid line.

ii. wavy: It is a wavy line.

iii. dotted: It is a dotted line.

iv. dashed: It is a line consisting of dashes.

v. double: It is a double solid line.

Here are some possible values under <text-decoration-color>

I. initial: It represents the value specified as the property's initial value.

II. inherit: It represents the calculated value of the property on the element's parent.

III. unset: It is a value that acts as either initial, depending on whether the property is inherited or not. This means that it sets all properties to the parent value if they are inheritable or to the initial value if not inheritable.

3. CSS text-decoration property.

Example:

```
<!DOCTYPE html>
<html>
<head>
<style>

.demo_container{
  width:600px;
  margin:0 auto;
  justify-content: center;
  align-items: center;
}
```

```
.div1 >h3{
 text-align: center;
 text-decoration: underline;
text-decoration-color: yellowgreen;
font-size: 28px;
}

.div2 > h3{
 text-align: center;
 text-decoration: underline;
 text-decoration-color: green;
 font-size: 28px;
}

.div3 >h3 {
  font-size: 28px;
  text-align: center;
  text-decoration-line: underline;
  text-decoration-style: wavy;
}

</style>
</head>
<body>

  <div class="demo_container">
    <h2> Seperate use of color, line, style
property of CSS text-decoration</h2>
    <div class="div1">
      <h3> CSS text-decoration-color property
</h3>
      <p class="text-1">
        Lorem ipsum sit amet consectetur
adipisicing elit. Maxime mollitia, molestiae
quas vel sint commodi repudiandae consequuntur
voluptatum laborum numquam blanditiis harum
quisquam eius sed odit fugiat iusto fuga
praesentium optio, eaque rerum!</h1>
      </p>
    </div>


    <div class=" div2">
      <h3>  CSS text-decoration-line
property  </h3>
```

```
    <p class="text-1">
        Lorem ipsum sit amet consectetur
adipisicing elit. Maxime mollitia, molestiae
quas vel sint commodi repudiandae consequuntur
voluptatum laborum numquam blanditiis harum
quisquam eius sed odit fugiat iusto fuga
praesentium optio, eaque rerum!</h1>
    </p>
    </div>


    <div class=" div3">
        <h3> CSS text-decoration-style
property  </h3>
        <p class="text-1">
        Lorem ipsum sit amet consectetur
adipisicing elit. Maxime mollitia, molestiae
quas vel sint commodi repudiandae consequuntur
voluptatum laborum numquam blanditiis harum
quisquam eius sed odit fugiat iusto fuga
praesentium optio, eaque rerum!</h1>
    </p>
    </div>
  </div>
</body>
</html>
```

The output of the code is given below:

Seperate use of color, line, style property of CSS text-decoration

CSS text-decoration-color property

Lorem ipsum sit amet consectetur adipisicing elit. Maxime mollitia, molestiae quas vel sint commodi repudiandae consequuntur voluptatum laborum numquam blanditiis harum quisquam eius sed odit fugiat iusto fuga praesentium optio, eaque rerum!

CSS text-decoration-line property

Lorem ipsum sit amet consectetur adipisicing elit. Maxime mollitia, molestiae quas vel sint commodi repudiandae consequuntur voluptatum laborum numquam blanditiis harum quisquam eius sed odit fugiat iusto fuga praesentium optio, eaque rerum!

CSS text-decoration-style property

Lorem ipsum sit amet consectetur adipisicing elit. Maxime mollitia, molestiae quas vel sint commodi repudiandae consequuntur voluptatum laborum numquam blanditiis harum quisquam eius sed odit fugiat iusto fuga praesentium optio, eaque rerum!

CSS text-decoration property.

4. CSS text-orientation property.

Example:

```
<!DOCTYPE html>
<html>
<head>
<style>

.demo_container{
  width:600px;
  margin:0 auto;
  justify-content: center;
  align-items: center;
}

p {
  writing-mode: vertical-rl;
  font-size: 2em;
  float: left;
}
.mixed {
  text-orientation: mixed;
  }
.upright {
  text-orientation: upright;
  }
.sideways {
  text-orientation: sideways;
  }

</style>
</head>
<body>

  <div class="demo_container">
    <h2> CSS text-orientation property </h2>
    <p class="mixed"> You are learning CSS </p>
    <p class="upright"> You are learning CSS </p>
    <p class="sideways"> You are learning CSS </p>
  </div>
</body>
</html>
```

The output of the code is given below:

CSS text-orientation property

You are learning CSS   C S S   Y o u   a r e   l e a r n i n g   You are learning CSS

CSS text-orientation property.

5. **CSS text-overflow property:** It specifies how text should be treated when it has been clipped due to it being too large to fit within its containing block.

Example:

```
<!DOCTYPE html>
<html>
<head>
<style>

.demo_container{
  width:600px;
  margin:0 auto;
  justify-content: center;
  align-items: center;
}

.text-1 {
  width: 16em;
  overflow: hidden;
  white-space: nowrap;
  background: gold;
}
```

```
.text-2{
  text-overflow: ellipsis;
}

</style>
</head>
<body>

  <div class="demo_container">
    <div class="div1">
      <h1> CSS text-overflow property </h1>

      <p class="text-1">
        Lorem ipsum sit amet consectetur
adipisicing elit. Maxime mollitia, molestiae
quas vel sint commodi repudiandae consequuntur
voluptatum laborum numquam blanditiis harum
quisquam eius sed odit fugiat iusto fuga
praesentium optio, eaque rerum!</h1>
      </p>
      <p class="text-2">
        Lorem ipsum sit amet consectetur
adipisicing elit. Maxime mollitia, molestiae
quas vel sint commodi repudiandae consequuntur
voluptatum laborum numquam blanditiis harum
quisquam eius sed odit fugiat iusto fuga
praesentium optio, eaque rerum!</h1>
      </p>
    </div>
  </div>

</body>
</html>
```

The output of the code is given below:

CSS text-overflow property

Lorem ipsum sit amet consectetur adipis

Lorem ipsum sit amet consectetur adipisicing elit. Maxime mollitia, molestiae quas vel sint commodi repudiandae consequuntur voluptatum laborum numquam blanditiis harum quisquam eius sed odit fugiat iusto fuga praesentium optio, eaque rerum!

CSS text-overflow property.

6. CSS text-shadow property: The text-shadow property is used for applying shadow effects to text. You can also use text-shadow to apply drop-shadows, outer glows, and other shadow effects to text. This property accepts a list of values. Each item in the list can have two, three, or four values.

Example:

```
<!DOCTYPE html>
<html>
<head>
<style>

.demo_container{
  width:600px;
  margin:0 auto;
  justify-content: center;
  align-items: center;
}

.text-1 {
  text-shadow: 2px 2px 5px red;
  font-size: 30px;

}
.text-2{
  text-shadow: 2px 2px red;
  font-size: 30px;
}
.text-3{
  text-shadow: 0 0 3px #ff0000, 0 0 5px #0000ff;
  font-size: 30px;
}

.text-4{
  color: white;
  text-shadow: 1px 1px 2px black, 0 0 25px blue,
0 0 5px darkblue;
  font-size: 30px;
}
</style>
</head>
<body>
```

```
<div class="demo_container">
  <div class="div1">
    <h1> CSS text-overflow property </h1>
    <p class="text-1">
      You are learning CSS.
  </p>
  <p class="text-2">
    You are learning CSS.
</p>
<p class="text-3">
  You are learning CSS.
</p>
<p class="text-4">
  You are learning CSS.
</p>
    </div>
  </div>

</body>
</html>
```

The output of the code is given below:

CSS text-shadow property

You are learning CSS.

You are learning CSS.

You are learning CSS.

You are learning CSS.

CSS text-shadow property.

CSS TRANSFORM PROPERTIES

The transform property is used to transform an element in two-dimensional (2D) or thre-dimensional (3D) spaces. For example, rotate elements, scale them, skew them, and more.

2D Transform Functions

The transform property accepts a list of "transform functions" as values. These functions have names such as scale(), rotate(), skew(), etc, which take parameters to determine the level of transformation (e.g., the angle to rotate an element).

The syntax of transform is given below:

```
transform: none | <transform-function> [ <transform-
function> ]
```

These are grouped by 2D (two-dimensional) and 3D (three-dimensional) functions.

matrix(): It specifies a 2D transformation in the form of a matrix of the six values a-f.

The syntax of matrix is given below:

```
matrix() = matrix( <number> [,<number> ]{5,5} )
```

translate(): It moves the position of the element. It specifies a 2D translation by the vector [x, y], where tx is the first translation-value parameter and y is the optional second translation-value parameter.

The syntax of translate is given below:

```
translate( <translation-value>[, <translation-value>]? )
```

translateX(): It moves the element horizontally. It specifies a translation by the given amount in the X direction.

The syntax of translateX is given below:

```
translateX( <translation-value> )
```

translateY(): It moves the element vertically. It specifies a translation by the amount in the Y direction.

The syntax of translateY is given below:

```
translateY( <translation-value> )
```

scale(): It modifies the size of the element. It specifies a 2D scale operation by the [sx,sy] vector described by two parameters.

The syntax of scale is given below:

```
scale( <number>[, <number>]? )
```

scaleX(): It specifies a 2D scale operation using the [sx,1] vector, where sx is given as the parameter.

The syntax of scaleX is given below:

```
scaleX( <number>[, <number>]? )
```

scaleY(): It specifies a 2D scale operation using the [1,sy] vector, where sy is given as the parameter.

The syntax of scaleY is given below:

```
scaleY( <number>[, <number>]? )
```

rotate(): It specifies a 2D rotation by the angle specified parameter about the origin, as defined by the transform-origin property.

The syntax of rotate is given below:

```
rotate( <angle> )
```

skew(): It specifies a 2D skew transformation along the X and Y axis by the given angles. If the second parameter is not provided, it has a zero value.

The syntax of skew is given below:

```
skew( <angle> [, <angle> ]? )
```

skewX(): It specifies a 2D transformation along the X-axis by the given angle.

The syntax of skewX is given below:

```
skewX( <angle> )
```

skewY(): It specifies a 2D transformation along the Y axis by the given angle.

The syntax of skewY is given below:

```
skewY( <angle> )
```

Example:

```
<!DOCTYPE html>
<html>
```

```
<head>
<style>

body{
  width: 90%;
  margin: 0 auto;
  font-family: sans-serif;
}
.wrap {
  display: inline-block;
  background: rgba(228, 225, 228, .6);
  width: 250px;
  height: 250px;
  margin: 60px 30px;
}
.box {
  border: 2px solid #444;
  line-height: 150px;
  font-size: .9em;
  font-weight: bold;
  text-align: center;
}
.box1 {
  background: red;
  transform: matrix(1, 0, 0, 1, 80, 80);
}
.box2 {
  background: #909;
  transform: translate(80px,80px);
}
.box3 {
  background: yellow;
  transform: matrix(.7, 0, 0, 1.3, 0, 0);
}
.box4 {
  background: green;
  transform: scale(.7, 1.3);
}
.box5 {
  background: pink;
  transform: matrix(1, 0, -0.5, 1, 0, 0);
}
```

```
.box6 {
  background: lightblue;
  transform: skew(-30deg);
}
.box7 {
  background: #f0f;
  transform: matrix(.71, -.71,.71, 0.71, 0, 0);
}
.box8 {
  background: #098;
  transform: rotate(-45deg);
}
</style>
</head>
<body>
<h1>2D Transform Functions</h1>
  <div class="wrap">
    <div class="box box1">matrix(1, 0, 0, 1, 80,
80)</div>
  </div>
  <div class="wrap">
    <div class="box box2">translate(80px,80px)
</div></div>
  <div class="wrap">
    <div class="box box3">matrix(.7, 0, 0, 1.3, 0,
0)</div>
  </div>
  <div class="wrap">
    <div class="box box4">scale(.7, 1.3)</div>
  </div>
  <div class="wrap">
    <div class="box box5">matrix(1, 0, -0.5, 1, 0,
0)</div>
  </div>
  <div class="wrap">
    <div class="box box6">skew(-30deg)</div>
  </div>
  <div class="wrap">
    <div class="box box7">matrix(.71, -.71,.71,
0.71, 0, 0);</div>
  </div>
  <div class="wrap">
```

```
    <div class="box box8">rotate(-45deg)</div>
    </div>

</body>
</html>
```

The output of the code is given below:

2D Transform Functions

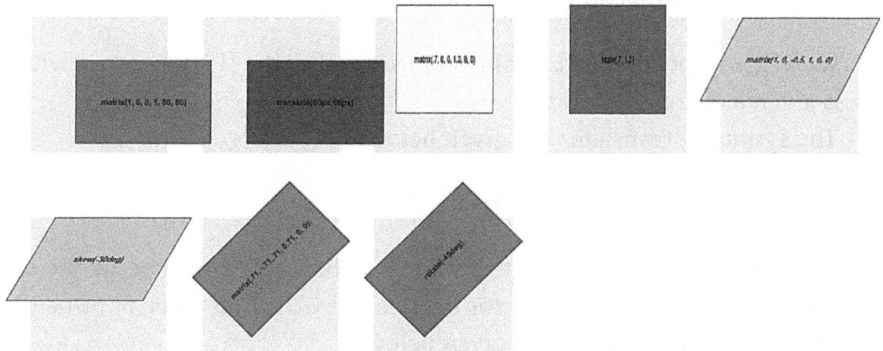

2D transform functions.

3D Transform Functions

It allows to apply transformations along three axes: x, y, and z axes, as demonstrated by a three-dimensional Cartesian coordinate system. The 3D transform functions available in CSS are given below.

matrix3d(): It specifies a 3D transformation as a 4x4 same matrix of 16 values in column-major order.

The syntax of matrix3d is given below:

```
matrix3d() = matrix3d( <number> [, <number> ]{15,15} )
```

translate3d(): It specifies a 3D translation by the vector [tx,ty,tz], tx, ty, and tz will be the first, second, and third translation value parameters, respectively.

The syntax of translate3d is given below:

```
translate3d() = translate3d( <translation-value>,
<translation-value>,  <length> )
```

translateZ(): It specifies a 3D translation by the vector [0,0,tz] with the given amount in the Z direction.

The syntax of translate3d is given below:

```
translateZ() = translateZ( <length> )
```

scale3d(): It specifies a 3D scale operation by the [sx,sy,sz] vector described by the three parameters.

```
scale3d() = scale3d( <number>,   <number>,   <number> )
```

scaleZ(): It specifies a 3D scale operation using the [1,1,sz] vector, where sz is given as the parameter.

The syntax of translate3d is given below:

```
scaleZ() = scaleZ( <number> )
```

rotate3d(): It specifies a 3D rotation by the angle specified in the last parameter about the [x,y,z] vector described by the first three parameters.

The syntax of translate3d is given below:

```
rotate3d() = rotate3d( <number>,   <number>,   <number>,
<number> )
```

rotateX(): This is the same as rotate3d(1, 0, 0, <angle>).

The syntax of rotateX is given below:

```
rotateX() = rotateX( <angle> )
```

rotateY(): This is the same as rotate3d(0, 1, 0, <angle>).

The syntax of rotateY is given below:

```
rotateY() = rotateY( <angle> )
```

rotateZ(): The same as rotate3d(0, 0, 1, <angle>) (which is also the same as rotate(<angle>).

The syntax of rotateZ is given below:

```
rotateZ() = rotateZ( <angle> )
```

perspective(): It defines the distance between the z=0 plane and the user in order to give to the 3D-positioned element some perspective.

The syntax of perspective is given below:

```
perspective() = perspective( <length> )
```

Example:

```
<!DOCTYPE html>
<html>
<head>
<style>

body { font-family: sans-serif; }

.scene {
  width: 200px;
  height: 200px;
  border: 1px solid #CCC;
  display: inline-block;
  width: 200px;
  height: 200px;
  margin: 60px 60px 60px 0;
  perspective: 600px;
}

.panel {
  width: 100%;
  height: 100%;
  background: hsla(0, 100%, 50%, 0.7);
  line-height: 200px;
  color: white;
  font-size: 18px;
  text-align: center;
}

.panel--translate-neg-z {
  transform: translateZ(-200px);
}

.panel--translate-pos-z {
  transform: translateZ(200px);
}
```

```
.panel--rotate-x {
  transform: rotateX(45deg);
}

.panel--rotate-y {
  transform: rotateY(45deg);
}

.panel--rotate-z {
  transform: rotateZ(45deg);
}
</style>
</head>
<body>
<h1> 3D Transform Functions</h1>
<div class="scene">
  <div class="panel panel--translate-neg-
z">translateZ(-200px)</div>
</div>

<div class="scene">
  <div class="panel panel--translate-pos-
z">translateZ(200px)</div>
</div>

<div class="scene">
  <div class="panel panel--rotate-
x">rotateX(45deg)</div>
</div>

<div class="scene">
  <div class="panel panel--rotate-
y">rotateY(45deg)</div>
</div>

<div class="scene">
  <div class="panel panel--rotate-
z">rotateZ(45deg)</div>
</div>
</body>
</html>
```

The output of the code is given below:

3D Transform Functions

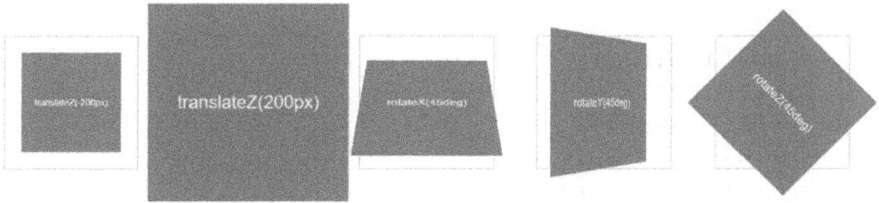

3D transform functions.

CSS TRANSITIONS PROPERTIES

CSS transition property is a property for defining CSS transitions that combines the four transition properties into one single property. The transition property combines the following properties such as transition-property, transition-duration, transition-timing-function, and transition-delay.

Example:

```
<!DOCTYPE html>
<html>
<head>
<style>

.demo_container{
  width:600px;
  margin:0 auto;
  justify-content: center;
  align-items: center;
}

.transition {
  background-color: yellow-green;
  color: RGB(77, 145, 213);
  width: 30%;
  height: 30px;
  padding: 10px;
  font-size: 1.5em;
  font-family: sans-serif;
```

```
    transition: width 0.5s cubic-bezier(0.25, 0.1,
0.25, 1.4) 0s, height 0.5s cubic-bezier(0.25, 0.1,
0.25, 1.4) 0s, font-size 0.5s cubic-bezier(0.25,
0.1, 0.25, 1.4) 0s;
}
.transition:hover {
  width: 60%;
  height: 180px;
  font-size: 2em;
  font-weight: bold;
  background-color: beige;
}
</style>
</head>
<body>

  <div class="demo_container">
    <div class="div1">
      <h1> CSS transition Properties </h1>
      <div class="transition"> Drage over me...
</div>
    </div>
  </div>

</body>
</html>
```

The output of the code is given below:

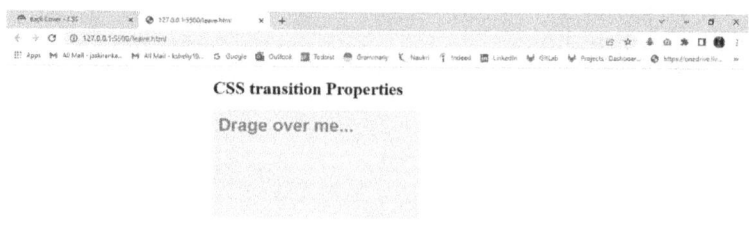

CSS transition properties.

CHAPTER SUMMARY

In this chapter, we have read about many properties of CSS in brief with examples so that it makes clear to understand the syntax of how to write code in CSS. The next chapter is about the most important concept in CSS named CSS Selectors.

CHAPTER 3

CSS Selectors

IN THIS CHAPTER

- ➤ Introduction

- ➤ Universal Selector

- ➤ Logical Combinations Selectors

- ➤ Attribute Selectors

- ➤ Pseudo-Classes

- ➤ Combinators Selectors

In the last chapter, we discussed various CSS properties, and now we have another important CSS: Selector. The selector means that the user can select any of the properties like classes, id, or tag names using the selector. The code can have various classes, but only unique id in it.

INTRODUCTION

Here is the list of the most common and well-supported CSS selectors. There are many more, but these are the ones you should know well.

- Type Selector: a { }

- ID Selector: #a { }

- Child Selector: a > b { }

- Descendant Selector: a b { }

DOI: 10.1201/9781003358060-3

- Selector list : a,b

- Combine Descendant & ID Selector: #a b { }

- Class Selector: .a { }

- Combine the Class Selector: b.x { }

- Comma Combinator Selector: a, c { }

- Universal Selector: * { }

- Combine Universal Selector: a * { }

- Adjacent Sibling Selector: a + b { }

- General Sibling Selector: a ~ b { }

- First Child Pseudo Selector: b:first-child { }

- Only Child Pseudo Selector: b:only-child { } or a :only-child { }

- Last Child Pseudo Selector: b:last-child { }

- Nth Child Pseudo Selector: a :nth-child(2) { }

- Nth Last Child Selector: :nth-last-child(2) { }

- First of Type Selector: b:first-of-type { }

- Nth of Type Selector: a:nth-of-type(2) { } or a:nth-of-type(even) { } or a:nth-of-type(odd) { } or a:nth-of-type(2n+1) { }

- Only of Type Selector: b:only-of-type { }

- Last of Type Selector: b:last-of-type { } or a :last-of-type { } or .x:last-of-type { }

- Empty Selector: a:empty { }

- Negation Pseudo-class Selector: a:not(.x) { } or a:not(:last-of-type) { }

- Attribute Selector: [for] { } or a[for] { }

- Attribute Value Selector: a[for="x"] { }

- Attribute Starts Selector: [for^="x"] { }

- Attribute Ends Selector: [for$="x"] { }

- Attribute Wildcard Selector: [for*="x"] { }

Let's have the example of each of the selectors with simple explanations, and the numbering can be changed from the above given selectors:

UNIVERSAL SELECTOR (*)

Example:

```
<!DOCTYPE html>
<html>
<head>
<style>

.demo_container{
  width:400px;
  margin:0 auto;
  text-align: center;
  justify-content: center;
  align-items: center;
}

* {
  background-color: #f13;
}
    </style>
</style>
</head>
<body>
<div class="demo_container">
  <h1> Universal Selectors (*) </h1>
  Lorem ipsum sit amet, consectetur adipiscing
elit.
   Aenean accumsan velit id lorem tempus iaculis.
   Praesent placerat lectus lorem. Quisque lacinia,
    metus sit amet tristique lacinia, nisi erat
pharetra lectus, in
     consequat turpis eros a erat. Cras blandit
vehicula arcu ac porta.
     Donec gravida massa vel odio pretium
dignissim. In id congue erat,
     quis fermentum nulla.
 </div>
  </body>
</html>
```

The output of the code is given below:

Universal Selectors (*)

Lorem ipsum dolor sit amet, consectetur adipiscing elit. Aenean accumsan velit id lorem tempus iaculis. Praesent placerat lectus lorem. Quisque lacinia, metus sit amet tristique lacinia, nisi erat pharetra lectus, in consequat turpis eros a erat. Cras blandit vehicula arcu ac porta. Donec gravida massa vel odio pretium dignissim. In id congue erat, quis fermentum nulla.

Universal selector (*).

UNIVERSAL SELECTOR (elements)

Example:

```
<!DOCTYPE html>
<html>
<head>
<style>

.demo_container{
  width:600px;
  margin:0 auto;
  text-align: center;
  justify-content: center;
  align-items: center;
}

div {
  background-color: #f13;

}
    </style>
</style>
</head>
```

```
<body>
<div class="demo_container">
  <h1> Universal Selectors (p, div, h1, aside,
section, article) </h1>
  Lorem ipsum sit amet, consectetur adipiscing
elit.
    Aenean accumsan velit id lorem tempus iaculis.
    Praesent placerat lectus lorem. Quisque lacinia,
    metus sit amet tristique lacinia, nisi erat
pharetra lectus, in
      consequat turpis eros a erat. Cras blandit
vehicula arcu ac porta.
      Donec gravida massa vel odio pretium
dignissim. In id congue erat,
      quis fermentum nulla.
 </div>
  </body>
</html>
```

The output of the code is given below:

Universal Selectors (p, div, h1, aside, section, article)

Lorem ipsum dolor sit amet, consectetur adipiscing elit. Aenean accumsan velit id lorem tempus iaculis. Praesent placerat lectus lorem. Quisque lacinia, metus sit amet tristique lacinia, nisi erat pharetra lectus, in consequat turpis eros a erat. Cras blandit vehicula arcu ac porta. Donec gravida massa vel odio pretium dignissim. In id congue erat, quis fermentum nulla.

Universal selector (elements/tags).

LOGICAL COMBINATIONS SELECTORS

LOGICAL COMBINATIONS SELECTORS (:not)

Example:

```
<!DOCTYPE html>
<html>
<head>
<style>
```

```
.demo_container{
  width:600px;
  margin:0 auto;
  text-align: center;
  justify-content: center;
  align-items: center;
}

div {
  background-color: #f13;
}
p:not(div,h1) {
background-color: yellow;
}
    </style>
</style>
</head>
<body>
<div class="demo_container">
  <h1> Logical Selectors ( :not ) </h1>
  Lorem ipsum sit amet, consectetur adipiscing
elit.
   Aenean accumsan velit id lorem tempus iaculis.
   Praesent placerat lectus lorem. Quisque
lacinia,
     <p>  metus sit amet tristique lacinia, nisi
erat pharetra lectus, in
       consequat turpis eros a erat. Cras blandit
vehicula arcu ac porta.
 </p>
 <p>   metus sit amet tristique lacinia, nisi erat
pharetra lectus, in
  consequat turpis eros a erat. Cras blandit
vehicula arcu ac porta.
</p>
</div>

  </body>
</html>
```

The output of the code is given below:

Logical Selectors (:not)

Lorem ipsum dolor sit amet, consectetur adipiscing elit. Aenean accumsan velit id lorem tempus iaculis. Praesent placerat lectus lorem. Quisque lacinia,

metus sit amet tristique lacinia, nisi erat pharetra lectus, in consequat turpis eros a erat. Cras blandit vehicula arcu ac porta.

metus sit amet tristique lacinia, nisi erat pharetra lectus, in consequat turpis eros a erat. Cras blandit vehicula arcu ac porta.

Logical selectors (:not).

ATTRIBUTE SELECTORS

ATTRIBUTE SELECTORS (.classname)

Example:

```
<!DOCTYPE html>
<html>
<head>
<style>

.demo_container{
  width:600px;
  margin:0 auto;
  text-align: center;
  justify-content: center;
  align-items: center;
}

p.p1{
background-color: yellow;
}
p.p2{
background-color: pink;
}
    </style>
</style>
</head>
```

```
<body>
<div class="demo_container">
  <h1> Attribute Selectors (.classname ) </h1>
  Lorem ipsum dolor sit amet, consectetur
adipiscing elit.
    Aenean accumsan velit id lorem tempus iaculis.
    Praesent placerat lectus lorem. Quisque
lacinia,
      <p class="p2">   metus sit amet tristique
lacinia, nisi erat pharetra lectus, in
      consequat turpis eros a erat. Cras blandit
vehicula arcu ac porta.
  </p>
  <p class="p1">   metus sit amet tristique
lacinia, nisi erat pharetra lectus, in
  consequat turpis eros a erat. Cras blandit
vehicula arcu ac porta.
  </p>
</div>

  </body>
</html>
```

The output of the code is given below:

Attribute Selectors (.classname)

Lorem ipsum dolor sit amet, consectetur adipiscing elit. Aenean accumsan velit id lorem tempus iaculis. Praesent placerat lectus lorem. Quisque lacinia,

metus sit amet tristique lacinia, nisi erat pharetra lectus, in consequat turpis eros a erat. Cras blandit vehicula arcu ac porta.

metus sit amet tristique lacinia, nisi erat pharetra lectus, in consequat turpis eros a erat. Cras blandit vehicula arcu ac porta.

Attribute selectors (.classname).

ATTRIBUTE SELECTORS (id)

Example:

```
<!DOCTYPE html>
<html>
```

```
<head>
<style>

.demo_container{
  width:600px;
  margin:0 auto;
  text-align: center;
  justify-content: center;
  align-items: center;
}

p#p1{
background-color: yellow;
}
p#p2{
background-color: pink;
}
    </style>
</style>
</head>
<body>
<div class="demo_container">
  <h1> Attribute Selectors (#id ) </h1>
  Lorem ipsum dolor sit amet, consectetur
adipiscing elit.
   Aenean accumsan velit id lorem tempus iaculis.
   Praesent placerat lectus lorem. Quisque
lacinia,
     <p id="p2">   metus sit amet tristique
lacinia, nisi erat pharetra lectus, in
     consequat turpis eros a erat. Cras blandit
vehicula arcu ac porta.
 </p>
 <p id="p1">   metus sit amet tristique lacinia,
nisi erat pharetra lectus, in
  consequat turpis eros a erat. Cras blandit
vehicula arcu ac porta.
</p>
</div>

  </body>
</html>
```

The output of the code is given below:

Attribute Selectors (#id)

Lorem ipsum dolor sit amet, consectetur adipiscing elit. Aenean accumsan velit id lorem tempus iaculis. Praesent placerat lectus lorem. Quisque lacinia,

metus sit amet tristique lacinia, nisi erat pharetra lectus, in consequat turpis eros a erat. Cras blandit vehicula arcu ac porta.

metus sit amet tristique lacinia, nisi erat pharetra lectus, in consequat turpis eros a erat. Cras blandit vehicula arcu ac porta.

Attribute selectors (#id).

ATTRIBUTE SELECTORS ([])

Example:

```
<!DOCTYPE html>
<html>
<head>
<style>

.demo_container{
  width:600px;
  margin:0 auto;
  text-align: center;
  justify-content: center;
  align-items: center;
}

abbr[title] {
  font-size: 24px;
}
    </style>
</style>
</head>
<body>
<div class="demo_container">
  <h1> Attribute Selectors ( [] ) </h1>
    <abbr title="This is title">
    Metus sit amet tristique lacinia, nisi erat
pharetra lectus, in
```

```
        consequat turpis eros a erat. Cras blandit
vehicula arcu ac porta.
  </abbr>

</div>

  </body>
</html>
```

The output of the code is given below:

Attribute Selectors ([])

Metus sit amet tristique lacinia, nisi erat pharetra lectus, in consequat turpis eros a erat. Cras blandit vehicula arcu ac porta.

Attribute selectors ([ ]).

ATTRIBUTE SELECTORS ([attr="value"])

Example:

```
<!DOCTYPE html>
<html>
<head>
<style>

.demo_container{
  width:600px;
  margin:0 auto;
  text-align: center;
  justify-content: center;
  align-items: center;
}

p[id="p1"]{
  font-size: 24px;
background-color: yellow;
}
```

```
p[id="p2"]{
background-color: pink;
}


    </style>
</style>
</head>
<body>
<div class="demo_container">
   <h1> Attribute Selectors ( [attr="value"] ) </h1>
   Lorem ipsum dolor sit amet, consectetur
adipiscing elit.
   Aenean accumsan velit id lorem tempus iaculis.
   Praesent placerat lectus lorem. Quisque lacinia,
      <p id="p2" >   metus sit amet tristique
lacinia, nisi erat pharetra lectus, in
      consequat turpis eros a erat. Cras blandit
vehicula arcu ac porta.
 </p>
 <p id="p1">   metus sit amet tristique lacinia,
nisi erat pharetra lectus, in
  consequat turpis eros a erat. Cras blandit
vehicula arcu ac porta.
</p>
</div>

   </body>
</html>
```

The output of the code is given below:

Attribute Selectors ([attr="value"])

Lorem ipsum dolor sit amet, consectetur adipiscing elit. Aenean accumsan velit id lorem tempus iaculis. Praesent placerat lectus lorem. Quisque lacinia,

metus sit amet tristique lacinia, nisi erat pharetra lectus, in consequat turpis eros a erat. Cras blandit vehicula arcu ac porta.

metus sit amet tristique lacinia, nisi erat pharetra lectus, in consequat turpis eros a erat. Cras blandit vehicula arcu ac porta.

Attribute selectors ([attr="value"]).

ATTRIBUTE SELECTORS ([attr="value"] case-sensitive)

Example:

```
<!DOCTYPE html>
<html>
<head>
<style>

.demo_container{
  width:600px;
  margin:0 auto;
  text-align: center;
  justify-content: center;
  align-items: center;
}

p[id="P1" i]{
  font-size: 24px;
background-color: yellow;
}

p[id="P2" i]{
background-color: pink;
}

    </style>
</style>
</head>
<body>
<div class="demo_container">
  <h1> Attribute Selectors ( [attr="value"] case-
sensitive) </h1>
  Lorem ipsum dolor sit amet, consectetur
adipiscing elit.
    Aenean accumsan velit id lorem tempus iaculis.
    Praesent placerat lectus lorem. Quisque
lacinia,
      <p id="p2" >   metus sit amet tristique
lacinia, nisi erat pharetra lectus, in
      consequat turpis eros a erat. Cras blandit
vehicula arcu ac porta.
  </p>
```

```
<p id="p1">   metus sit amet tristique lacinia,
nisi erat pharetra lectus, in
  consequat turpis eros a erat. Cras blandit
vehicula arcu ac porta.
</p>
</div>

  </body>
</html>
```

The output of the code is given below:

Attribute Selectors ([attr="value"] case-sensitive)

Lorem ipsum dolor sit amet, consectetur adipiscing elit. Aenean accumsan velit id lorem tempus iaculis. Praesent placerat lectus lorem. Quisque lacinia,

metus sit amet tristique lacinia, nisi erat pharetra lectus, in consequat turpis eros a erat. Cras blandit vehicula arcu ac porta.

metus sit amet tristique lacinia, nisi erat pharetra lectus, in consequat turpis eros a erat. Cras blandit vehicula arcu ac porta.

Attribute selectors ([attr="value"] case-sensitive).

ATTRIBUTE SELECTORS ([foo~="bar"])

Example:

```
<!DOCTYPE html>
<html>
<head>
<style>

.demo_container{
  width:600px;
  margin:0 auto;
  text-align: center;
  justify-content: center;
  align-items: center;
}
```

```
[class~="ABC"] {
    color: orange;
}



    </style>
</style>
</head>
<body>
<div class="demo_container">
  <h1> Attribute Selectors ( [ foo~="bar" ] ) </h1>
  Lorem ipsum dolor sit amet, consectetur
adipiscing elit.
    Aenean accumsan velit id lorem tempus iaculis.
    Praesent placerat lectus lorem. Quisque
lacinia,
 <section class="abc">  metus sit amet tristique
lacinia, nisi erat pharetra lectus, in
  consequat turpis eros a erat. Cras blandit
vehicula arcu ac porta.
</section>
<p class="abc">  metus sit amet tristique
lacinia, nisi erat pharetra lectus, in
  consequat turpis eros a erat. Cras blandit
vehicula arcu ac porta.
</p>
</div>

    </body>
</html>
```

The output of the code is given below:

Attribute Selectors ([foo~="bar"])

Lorem ipsum dolor sit amet, consectetur adipiscing elit. Aenean accumsan velit id lorem tempus iaculis. Praesent placerat lectus lorem. Quisque lacinia,

metus sit amet tristique lacinia, nisi erat pharetra lectus, in consequat turpis eros a erat. Cras blandit vehicula arcu ac porta.

metus sit amet tristique lacinia, nisi erat pharetra lectus, in consequat turpis eros a erat. Cras blandit vehicula arcu ac porta.

Attribute selectors ([foo~="bar"]).

ATTRIBUTE SELECTORS ([foo^="bar"])

Example:

```
<!DOCTYPE html>
<html>
<head>
<style>

.demo_container{
  width:600px;
  margin:0 auto;
  text-align: center;
  justify-content: center;
  align-items: center;
}

[class^="title"] {
  background-color: orchid;
    font-size: 20px;
}


    </style>
</style>
</head>
<body>
<div class="demo_container">
  <h1> Attribute Selectors ( [ foo^="bar" ] ) </h1>
  Lorem ipsum dolor sit amet, consectetur
adipiscing elit.
   Aenean accumsan velit id lorem tempus
iaculis.
   Praesent placerat lectus lorem. Quisque
lacinia,
 <section class="title-1">  metus sit amet
tristique lacinia, nisi erat pharetra lectus, in
  consequat turpis eros a erat. Cras blandit
vehicula arcu ac porta.
 </section>
```

```
<p class="title-2">   metus sit amet tristique
lacinia, nisi erat pharetra lectus, in
  consequat turpis eros a erat. Cras blandit
vehicula arcu ac porta.
</p>
</div>

  </body>
</html>
```

The output of the code is given below:

Attribute Selectors ([foo^="bar"])

Lorem ipsum dolor sit amet, consectetur adipiscing elit. Aenean accumsan velit id lorem
tempus iaculis. Praesent placerat lectus lorem. Quisque lacinia,

metus sit amet tristique lacinia, nisi erat pharetra lectus, in consequat
turpis eros a erat. Cras blandit vehicula arcu ac porta.

metus sit amet tristique lacinia, nisi erat pharetra lectus, in consequat
turpis eros a erat. Cras blandit vehicula arcu ac porta.

Attribute selectors ([foo^= "bar"]).

ATTRIBUTE SELECTORS ([foo$="bar"])

Example:

```
<!DOCTYPE html>
<html>
<head>
<style>

.demo_container{
  width:600px;
  margin:0 auto;
  justify-content: center;
  align-items: center;
}

[class$="one"] {
  background-color: orchid;
}
```

```
[class$="two"] {
  background-color: orange;
}



    </style>
</style>
</head>
<body>
<div class="demo_container">
  <h1> Attribute Selectors ( [ foo$="bar" ] ) </h1>
    <ul>    <h1> List 1 </h1>
    <li class="list-one">   Metus sit amet
tristique lacinia, nisi erat pharetra lectus, in
    consequat turpis eros a erat. Cras blandit
vehicula arcu ac porta.
    </li>
    <li class="list-one">   Metus sit amet
tristique lacinia, nisi erat pharetra lectus, in
    consequat turpis eros a erat. Cras blandit
vehicula arcu ac porta.
    </li>
    <li class="list-one">   Metus sit amet
tristique lacinia, nisi erat pharetra lectus, in
    consequat turpis eros a erat. Cras blandit
vehicula arcu ac porta.
    </li>
    </ul>


    <ul>    <h2> List 2 </h2>
    <li class="list-two">   Metus sit amet
tristique lacinia, nisi erat pharetra lectus, in
    consequat turpis eros a erat. Cras blandit
vehicula arcu ac porta.
    </li>
    <li class="list-two">   Metus sit amet
tristique lacinia, nisi erat pharetra lectus, in
    consequat turpis eros a erat. Cras blandit
vehicula arcu ac porta.
    </li>
```

```
    <li class="list-two">   Metus sit amet
tristique lacinia, nisi erat pharetra lectus, in
     consequat turpis eros a erat. Cras blandit
vehicula arcu ac porta.
    </li>
    </ul>
  </div>

    </body>
  </html>
```

The output of the code is given below:

Attribute Selectors ([foo$="bar"])

List 1

- Metus sit amet tristique lacinia, nisi erat pharetra lectus, in consequat turpis eros a erat. Cras blandit vehicula arcu ac porta.
- Metus sit amet tristique lacinia, nisi erat pharetra lectus, in consequat turpis eros a erat. Cras blandit vehicula arcu ac porta.
- Metus sit amet tristique lacinia, nisi erat pharetra lectus, in consequat turpis eros a erat. Cras blandit vehicula arcu ac porta.

List 2

- Metus sit amet tristique lacinia, nisi erat pharetra lectus, in consequat turpis eros a erat. Cras blandit vehicula arcu ac porta.
- Metus sit amet tristique lacinia, nisi erat pharetra lectus, in consequat turpis eros a erat. Cras blandit vehicula arcu ac porta.
- Metus sit amet tristique lacinia, nisi erat pharetra lectus, in consequat turpis eros a erat. Cras blandit vehicula arcu ac porta.

Attribute selectors ([foo$= "bar"]).

ATTRIBUTE SELECTORS ([foo*="bar"])

Example:

```
<!DOCTYPE html>
<html>
<head>
<style>
```

```
.demo_container{
  width:600px;
  margin:0 auto;
  justify-content: center;
  align-items: center;
}

[class*="one"] {
  color: orchid;
  background-color: black;
}

[class$="two"] {
  color: orange;
  background-color: black;

}

</style>
</style>
</head>
<body>
<div class="demo_container">
  <h1> Attribute Selectors ( [ foo*="bar" ] ) </h1>
    <ul>   <h1>  List 1 </h1>
     <li class="list-one">  Metus sit amet
tristique lacinia, nisi erat pharetra lectus, in
     consequat turpis eros a erat. Cras blandit
vehicula arcu ac porta.
     </li>
     <li class="list-one">  Metus sit amet
tristique lacinia, nisi erat pharetra lectus, in
     consequat turpis eros a erat. Cras blandit
vehicula arcu ac porta.
     </li>
     <li class="list-one">  Metus sit amet
tristique lacinia, nisi erat pharetra lectus, in
     consequat turpis eros a erat. Cras blandit
vehicula arcu ac porta.
     </li>
    </ul>
```

```
<ul>    <h2> List 2 </h2>
    <li class="list-two">   Metus sit amet
tristique lacinia, nisi erat pharetra lectus, in
    consequat turpis eros a erat. Cras blandit
vehicula arcu ac porta.
    </li>
    <li class="list-two">   Metus sit amet
tristique lacinia, nisi erat pharetra lectus, in
    consequat turpis eros a erat. Cras blandit
vehicula arcu ac porta.
    </li>
    <li class="list-two">   Metus sit amet
tristique lacinia, nisi erat pharetra lectus, in
    consequat turpis eros a erat. Cras blandit
vehicula arcu ac porta.
    </li>
    </ul>
  </div>

  </body>
</html>
```

The output of the code is given below:

Attribute Selectors ([foo*="bar"])

List 1

- Metus sit amet tristique lacinia, nisi erat pharetra lectus, in consequat turpis eros a erat. Cras blandit vehicula arcu ac porta.
- Metus sit amet tristique lacinia, nisi erat pharetra lectus, in consequat turpis eros a erat. Cras blandit vehicula arcu ac porta.
- Metus sit amet tristique lacinia, nisi erat pharetra lectus, in consequat turpis eros a erat. Cras blandit vehicula arcu ac porta.

List 2

- Metus sit amet tristique lacinia, nisi erat pharetra lectus, in consequat turpis eros a erat. Cras blandit vehicula arcu ac porta.
- Metus sit amet tristique lacinia, nisi erat pharetra lectus, in consequat turpis eros a erat. Cras blandit vehicula arcu ac porta.
- Metus sit amet tristique lacinia, nisi erat pharetra lectus, in consequat turpis eros a erat. Cras blandit vehicula arcu ac porta.

Attribute selectors ([foo*="bar"]).

ATTRIBUTE SELECTORS ([foo|="bar"])

Example:

```
<!DOCTYPE html>
<html>
<head>
<style>

.demo_container{
  width:600px;
  margin:0 auto;
  justify-content: center;
  align-items: center;
}



[class|="list"] {
  color: orange;
  background-color: black;

}

</style>
</style>
</head>
<body>
<div class="demo_container">
  <h1> Attribute Selectors ( [ foo|="bar" ] ) </h1>
    <ul>   <h1>  List 1 </h1>
     <li class="list-one">   Metus sit amet
tristique lacinia, nisi erat pharetra lectus, in
     consequat turpis eros a erat. Cras blandit
vehicula arcu ac porta.
     </li>
     <li class="list-one">   Metus sit amet
tristique lacinia, nisi erat pharetra lectus, in
     consequat turpis eros a erat. Cras blandit
vehicula arcu ac porta.
     </li>
```

```
    <li class="list-one">   Metus sit amet
tristique lacinia, nisi erat pharetra lectus, in
    consequat turpis eros a erat. Cras blandit
vehicula arcu ac porta.
    </li>
    </ul>

  </div>

  </body>
</html>
```

The output of the code is given below:

Attribute Selectors ([foo|="bar"])

List 1

Attribute selectors ([foo|="bar"]).

PSEUDO-CLASSES

PSEUDO-CLASSES (:dir(ltr) or :dir(rtl))

Example:

```
<!DOCTYPE html>
<html>
<head>
<style>

.demo_container{
  width:600px;
  margin:0 auto;
```

```
    justify-content: center;
    align-items: center;
}


div:dir(ltr) {
   background-color: #333;
   color: #fff;
}

div:dir(rtl) {
   background: red;
   color: #fff;
}

</style>
</style>
</head>
<body>
<div class="demo_container">
   <h1>    Pseudo-Classes (:dir(ltr) or :dir(rtl))
</h1>

<div dir="ltr">
  <h2>left-to-right</h2>
  <p> Metus sit amet tristique lacinia, nisi erat
pharetra lectus, in
      consequat turpis eros a erat. Cras blandit
vehicula arcu ac porta.</p>
  <cite>&mdash; Lorem
</div>
<div dir="rtl">
  <h1>right to left</h1>
  <p> Metus sit amet tristique lacinia, nisi erat
pharetra lectus, in
      consequat turpis eros a erat. Cras blandit
vehicula arcu ac porta. </p>
  <cite>&mdash;Lorem</cite>
</div>
 </div>
  </body>
</html>
```

The output of the code is given below:

Pseudo-Classes (:dir(ltr) or :dir(rtl))

left-to-right

Metus sit amet tristique lacinia, nisi erat pharetra lectus, in consequat turpis eros a erat. Cras blandit vehicula arcu ac porta.

— *Lorem*

right to left

Metus sit amet tristique lacinia, nisi erat pharetra lectus, in consequat turpis eros a erat. Cras .blandit vehicula arcu ac porta

Lorem—

Pseudo-classes (:dir(ltr) or :dir(rtl)).

PSEUDO-CLASSES (:any-link)

Example:

```
<!DOCTYPE html>
<html>
<head>
<style>

.demo_container{
  width:600px;
  margin:0 auto;
  justify-content: center;
  align-items: center;
}

a:any-link {
        background-color: green;
        color: white;
        text-decoration:none
    }
```

```
a:hover{
  font-size: 17px;
}

</style>
</style>
</head>
<body>
<div class="demo_container">
  <h1>   Pseudo-Classes (:any-link) </h1>
<p> In the below paragraph you will get
highlighted text that text is the any link present
in the paragraph </p>
<div dir="ltr">
  <p>
    Lorem ipsum sit amet, consectetur adipiscing
elit.
    Mauris convallis ipsum vitae nibh venenatis,
id lacinia dui ultricies.
    Pellentesque feugiat quis ante quis accumsan.
Nulla ut mi quis felis eleifend semper.
    Mauris quis magna neque. Maecenas non velit
cursus, sagittis orci eget, viverra neque.
    <a href="#">Nullam mollis ornare mauris, id
finibus nunc fermentum eget. Fusce efficitur
luctus volutpat. </a>
    Orci varius natoque et magnis dis parturient
montes, nascetur ridiculus mus.
    Aliquam id tristique urna, sed gravida massa.
Nam sed magna mi. Maecenas rhoncus nisi ipsum,
    suscipit ultricies mi volutpat in. Nullam non
odio nec sapien posuere rutrum vitae at dolor.
    <a href="#">Etiam in tristique justo, a
posuere neque. Donec sed pulvinar lacus.
Vestibulum at nunc enim. </a>
    Aenean id facilisis erat, a pulvinar lectus.
  </p>

<br>
</div>
</body>
</html>
```

The output of the code is given below:

Pseudo-Classes (:any-link)

In the below paragraph you will get highlighted text that text is the any link present in the paragraph

Lorem ipsum dolor sit amet, consectetur adipiscing elit. Mauris convallis ipsum vitae nibh venenatis, id lacinia dui ultricies. Pellentesque feugiat quis ante quis accumsan. Nulla ut mi quis felis eleifend semper. Mauris quis magna neque. Maecenas non velit cursus, sagittis orci eget, viverra neque. Nullam mollis ornare mauris, id finibus nunc fermentum eget. Fusce efficitur luctus volutpat. Orci varius natoque penatibus et magnis dis parturient montes, nascetur ridiculus mus. Aliquam id tristique urna, sed gravida massa. Nam sed magna mi. Maecenas rhoncus nisi ipsum, suscipit ultricies mi volutpat in. Nullam non odio nec sapien posuere rutrum vitae at dolor. Etiam in tristique justo, a posuere neque. Donec sed pulvinar lacus. Vestibulum at nunc enim. Aenean id facilisis erat, a pulvinar lectus.

Pseudo-classes (:any-link).

PSEUDO-CLASSES (:link)

Example:

```
<!DOCTYPE html>
<html>
<head>
<style>

.demo_container{
  width:600px;
  margin:0 auto;
  justify-content: center;
  align-items: center;
}

a:link {
    color: orange;
}
```

```
a:hover{
  font-size: 17px;
}

</style>
</style>
</head>
<body>
<div class="demo_container">
  <h1>   Pseudo-Classes (:link) </h1>
<p> In the below paragraph you will get
highlighted text, that is the link which is not
visited yet </p>
<div dir="ltr">
  <p>
    Lorem ipsum dolor sit amet, consectetur
adipiscing elit.
    Mauris convallis ipsum vitae nibh venenatis,
id lacinia dui ultricies.
    Pellentesque feugiat quis ante quis accumsan.
Nulla ut mi quis felis eleifend semper.
    Mauris quis magna neque. Maecenas non velit
cursus, sagittis orci eget, viverra neque.
    <a href="#">Nullam mollis ornare mauris, id
finibus nunc fermentum eget. Fusce efficitur
luctus volutpat. </a>
    Orci varius natoque et magnis dis parturient
montes, nascetur ridiculus mus.
    Aliquam id tristique urna, sed gravida massa.
Nam sed magna mi. Maecenas rhoncus nisi ipsum,
    suscipit ultricies mi volutpat in. Nullam non
odio nec sapien posuere rutrum vitae at dolor.
    <a href="#">Etiam in tristique justo, a
posuere neque. Donec sed pulvinar lacus.
Vestibulum at nunc enim. </a>
    Aenean id facilisis erat, a pulvinar lectus.
  </p>

<br>
</div>
</body>
</html>
```

The output of the code is given below:

Pseudo-Classes (:link)

In the below paragraph you will get highlighted text, that is the link which is not visited yet

Lorem ipsum dolor sit amet, consectetur adipiscing elit. Mauris convallis ipsum vitae nibh venenatis, id lacinia dui ultricies. Pellentesque feugiat quis ante quis accumsan. Nulla ut mi quis felis eleifend semper. Mauris quis magna neque. Maecenas non velit cursus, sagittis orci eget, viverra neque. Nullam mollis ornare mauris, id finibus nunc fermentum eget. Fusce efficitur luctus volutpat. Orci varius natoque penatibus et magnis dis parturient montes, nascetur ridiculus mus. Aliquam id tristique urna, sed gravida massa. Nam sed magna mi. Maecenas rhoncus nisi ipsum, suscipit ultricies mi volutpat in. Nullam non odio nec sapien posuere rutrum vitae at dolor. Etiam in tristique justo, a posuere neque. Donec sed pulvinar lacus. Vestibulum at nunc enim. Aenean id facilisis erat, a pulvinar lectus.

Pseudo-classes (:link).

PSEUDO-CLASSES (:target)

Example:

```
<!DOCTYPE html>
<html>
<head>
<style>

.demo_container{
  width:600px;
  margin:0 auto;
  justify-content: center;
  align-items: center;
}

:target {
  border: 2px solid #D4D4D4;
  background-color: #f1f;
}


</style>
</style>
</head>
```

```
<body>
<div class="demo_container">
  <h1>   Pseudo-Classes (:target) </h1>
<p> In the below paragraph you will get
highlighted text, link which will target. </p>
<div dir="ltr">
  <p><a href="#link1"> Move to Link 1</a></p>
<p><a href="#link2"> Move to Link 2</a></p>
  <p>
    Lorem ipsum dolor sit amet, consectetur
adipiscing elit.
    Mauris convallis ipsum vitae nibh venenatis,
id lacinia dui ultricies.
    Pellentesque feugiat quis ante quis accumsan.
Nulla ut mi quis felis eleifend semper.
    Mauris quis magna neque. Maecenas non velit
cursus, sagittis orci eget, viverra neque.
      <a href="#"  id="link1">Nullam mollis ornare
mauris, id finibus nunc fermentum eget. Fusce
efficitur luctus volutpat. </a>
    Orci varius natoque penatibus et magnis dis
parturient montes, nascetur ridiculus mus.
    Aliquam id tristique urna, sed gravida
massa. Nam sed magna mi. Maecenas rhoncus nisi
ipsum,
    suscipit ultricies mi volutpat in. Nullam
non odio nec sapien posuere rutrum vitae at
dolor.
      <a href="#"  id="link2">Etiam in tristique
justo, a posuere neque. Donec sed pulvinar
lacus. Vestibulum at nunc enim. </a>
    Aenean id facilisis erat, a pulvinar
lectus.
  </p>

  <br>
</div>
</body>
</html>
```

The output of the code is given below:

Pseudo-Classes (:target)

In the below paragraph you will get highlighted text, link which will target.

Move to Link 1

Move to Link 2

Lorem ipsum dolor sit amet, consectetur adipiscing elit. Mauris convallis ipsum vitae nibh venenatis, id lacinia dui ultricies. Pellentesque feugiat quis ante quis accumsan. Nulla ut mi quis felis eleifend semper. Mauris quis magna neque. Maecenas non velit cursus, sagittis orci eget, viverra neque. Nullam mollis ornare mauris, id finibus nunc fermentum eget. Fusce efficitur luctus volutpat. Orci varius natoque penatibus et magnis dis parturient montes, nascetur ridiculus mus. Aliquam id tristique urna, sed gravida massa. Nam sed magna mi. Maecenas rhoncus nisi ipsum, suscipit ultricies mi volutpat in. Nullam non odio nec sapien posuere rutrum vitae at dolor. Etiam in tristique justo, a posuere neque. Donec sed pulvinar lacus. Vestibulum at nunc enim. Aenean id facilisis erat, a pulvinar lectus.

Pseudo-classes (:target).

PSEUDO-CLASSES (:scope)

Example:

```
<!DOCTYPE html>
<html>
  <head>
    <title>Title of the document</title>
    <style>
      .container {
        margin: 40px auto;
        max-width: 700px;
        background-color: #eeeeee;
        padding: 20px;
        box-shadow: 0 0 3px rgba(0, 0, 0, 0.25);
      }
      section {
        padding: 30px;
      }
      :scope {
        background-color: #1c87c9;
      }
    </style>
  </head>
```

```
<body>
  <h2> Pseudo-Classes (:scope) </h2>
  <div class="container">
    <section>
      <p>
        Inside the scope.
      </p>
    </section>
  </div>
</body>
</html>
```

The output of the code is given below:

Pseudo-classes (:scope).

PSEUDO-CLASSES (:target)

Example:

```
<!DOCTYPE html>
<html>
<head>
<style>

.demo_container{
  width:600px;
  margin:0 auto;
  justify-content: center;
  align-items: center;
}
```

```
a:active {
  background-color: yellow;
}



</style>
</style>
</head>
<body>
<div class="demo_container">
  <h1>   Pseudo-Classes (:target) </h1>
<p> In the below links you will get highlighted
links, when clicked </p>
<div dir="ltr">
  <p><a href="#link1"> Move to Link 1</a></p>
<p><a href="#link2"> Move to Link 2</a></p>
  <p>
    Lorem ipsum dolor sit amet, consectetur
adipiscing elit.
    Mauris convallis ipsum vitae nibh venenatis,
id lacinia dui ultricies.
    Pellentesque feugiat quis ante quis accumsan.
Nulla ut mi quis felis eleifend semper.
    Mauris quis magna neque. Maecenas non velit
cursus, sagittis orci eget, viverra neque.
    <a href="#"  id="link1">Nullam mollis ornare
mauris, id finibus nunc fermentum eget. Fusce
efficitur luctus volutpat. </a>
    Orci varius natoque penatibus et magnis dis
parturient montes, nascetur ridiculus mus.
    Aliquam id tristique urna, sed gravida massa.
Nam sed magna mi. Maecenas rhoncus nisi ipsum,
    suscipit ultricies mi volutpat in. Nullam non
odio nec sapien posuere rutrum vitae at dolor.
    <a href="#"  id="link2">Etiam in tristique
justo, a posuere neque. Donec sed pulvinar lacus.
Vestibulum at nunc enim. </a>
    Aenean id facilisis erat, a pulvinar lectus.
  </p>

<br>
</div>
</body>
</html>
```

The output of the code is given below:

Pseudo-Classes (:target)

In the below links you will get highlighted links, when clicked

Move to Link 1

Move to Link 2

Lorem ipsum dolor sit amet, consectetur adipiscing elit. Mauris convallis ipsum vitae nibh venenatis, id lacinia dui ultricies. Pellentesque feugiat quis ante quis accumsan. Nulla ut mi quis felis eleifend semper. Mauris quis magna neque. Maecenas non velit cursus, sagittis orci eget, viverra neque. Nullam mollis ornare mauris, id finibus nunc fermentum eget. Fusce efficitur luctus volutpat. Orci varius natoque penatibus et magnis dis parturient montes, nascetur ridiculus mus. Aliquam id tristique urna, sed gravida massa. Nam sed magna mi. Maecenas rhoncus nisi ipsum, suscipit ultricies mi volutpat in. Nullam non odio nec sapien posuere rutrum vitae at dolor. Etiam in tristique justo, a posuere neque. Donec sed pulvinar lacus. Vestibulum at nunc enim. Aenean id facilisis erat, a pulvinar lectus.

Pseudo-classes (:target).

PSEUDO-CLASSES (:hover)

Example:

```
<!DOCTYPE html>
<html>
<head>
<style>

.demo_container{
  width:600px;
  margin:0 auto;
  justify-content: center;
  align-items: center;
}

a {
    color: red;
    text-decoration: none;
}
```

```
a:hover{
    background-color:black;
    font-size: 17px;
}


</style>
</style>
</head>
<body>
<div class="demo_container">
    <h1>   Pseudo-Classes (:hover) </h1>
<div div="text">
    <h3> Drag your mouse over the text which is in
red color </h3>
    <p>
        Lorem ipsum dolor sit amet, consectetur
adipiscing elit.
        Mauris convallis ipsum vitae nibh venenatis,
id lacinia dui ultricies.
        Pellentesque feugiat quis ante quis accumsan.
Nulla ut mi quis felis eleifend semper.
        Mauris quis magna neque. Maecenas non velit
cursus, sagittis orci eget, viverra neque.
        <a href="#">Nullam mollis ornare mauris, id
finibus nunc fermentum eget. Fusce efficitur
luctus volutpat. </a>
        Orci varius natoque penatibus et magnis dis
parturient montes, nascetur ridiculus mus.
        Aliquam id tristique urna, sed gravida massa.
Nam sed magna mi. Maecenas rhoncus nisi ipsum,
        suscipit ultricies mi volutpat in. Nullam non
odio nec sapien posuere rutrum vitae at dolor.
        <a href="#">Etiam in tristique justo, a
posuere neque. Donec sed pulvinar lacus.
Vestibulum at nunc enim. </a>
        Aenean id facilisis erat, a pulvinar lectus.
    </p>

<br>
</div>
</body>
</html>
```

The output of the code is given below:

Pseudo-Classes (:hover)

Drag your mouse over the text which is in red color

Lorem ipsum dolor sit amet, consectetur adipiscing elit. Mauris convallis ipsum vitae nibh venenatis, id lacinia dui ultricies. Pellentesque feugiat quis ante quis accumsan. Nulla ut mi quis felis eleifend semper. Mauris quis magna neque. Maecenas non velit cursus, sagittis orci eget, viverra neque. Nullam mollis ornare mauris, id finibus nunc fermentum eget. Fusce efficitur luctus volutpat. Orci varius natoque penatibus et magnis dis parturient montes, nascetur ridiculus mus. Aliquam id tristique urna, sed gravida massa. Nam sed magna mi. Maecenas rhoncus nisi ipsum, suscipit ultricies mi volutpat in. Nullam non odio nec sapien posuere rutrum vitae at dolor. Etiam in tristique justo, a posuere neque. Donec sed pulvinar lacus. Vestibulum at nunc enim. Aenean id facilisis erat, a pulvinar lectus.

Pseudo classes-7.png.

PSEUDO-CLASSES (:focus)

Example:

```
<!DOCTYPE html>
<html>
<head>
<style>

.demo_container{
  width:600px;
  margin:0 auto;
  justify-content: center;
  align-items: center;
}


label, input {
  width: 70%;
  padding: 0.5rem;
  box-sizing: border-box;
  justify-content: space-between;
  font-size: 1.1rem;
}
```

```
input:focus {
  background-color: lightblue;
}
label {
  text-align: right;
  width: 30%;
}
input {
  border: 2px solid #aaa;
  border-radius: 2px;
}

</style>
</head>
<body>
<div class="demo_container">
  <h1> Pseudo-Classes (:focus)  </h1>
  <form method="post" action="/" id="form"
class="validate">
    <div class="form-field">
      <label for="full-name">Full Name</label>
      <input type="text" name="full-name"
id="full-name" placeholder="John" required />
    </div> <br>
    <div class="form-field">
      <label for="email-input">Email</label>
      <input type="email" name="email-input"
id="email-input" placeholder="abc@gmail.com"
required />
    </div>
  </form>
</div>
</body>
</html>
```

The output of the code is given below:

Pseudo-Classes (:focus)

Full Name | John |

Email | abc@gmail.com |

Pseudo-classes-8.png.

PSEUDO-CLASSES (:enabled and :disabled)

Example:

```
<!DOCTYPE html>
<html>
<head>
<style>

.demo_container{
  width:600px;
  margin:0 auto;
  justify-content: center;
  align-items: center;
}



label, input {
  width: 70%;
  padding: 0.5rem;
  box-sizing: border-box;
  justify-content: space-between;
  font-size: 1.1rem;
}
input[type="text"]:enabled {
  background-color: lightblue;
}
input[type="email"]:disabled {
  background-color: lightgray;
}
label {
  text-align: right;
  width: 30%;
}
input {
  border: 2px solid #aaa;
  border-radius: 2px;
}

</style>
</head>
```

```
<body>
<div class="demo_container">
   <h1> Pseudo-Classes (:enabled and :disabled)  </h1>
   <h3> In the form the field having type text is
enabled and email is disabled </h3> <br>
   <form method="post" action="/" id="form"
class="validate">
     <div class="form-field">
        <label for="full-name">Full Name</label>
        <input type="text" name="full-name"
id="full-name" placeholder="John" required />
     </div> <br> <br>
     <div class="form-field">
        <label for="email-input">Email</label>
        <input type="email" name="email-input"
disabled="disabled" id="email-input"
placeholder="abc@gmail.com" required />
     </div>
   </form>
</div>
</body>
</html>
```

The output of the code is given below:

Pseudo-Classes (:focus)

In the form the field having type text is enabled and email is disabled

Full Name John

Email abc@gmail.com

Pseudo-classes (:enabled and :disabled).

PSEUDO-CLASSES (:read-only and :read-write)

Example:

```
<!DOCTYPE html>
<html>
<head>
<style>
```

```
.demo_container{
  width:600px;
  margin:0 auto;
  justify-content: center;
  align-items: center;
}


@import "compass/css3";

.test {
  display: block;
  margin-bottom: 1em;
  border: 1px solid silver;
  width: 100%;
  padding: .5em;
}

.test:read-only {
  background: tomato;
}

.test:read-write {
  background: light green;
}

* {
  box-sizing: border-box;
}

body {
  padding: 1em;
}
</style>
</head>
<body>
<div class="demo_container">
  <h1> Pseudo-Classes (:read-only and :read-
write)  </h1>
  <h3> In the form the field having type text is
enabled and email is disabled </h3> <br>
```

```
<form method="post" action="/" id="form"
class="validate">
    <input class="test" type="text" value="Regular
input" /> <br>
    <input class="test" type="text"
value="Disabled input" disabled /> <br>
    <input class="test" type="text"
value="Readonly input" readonly /> <br>
    <p class="test"
contenteditable>Contenteditable paragraph</p> <br>
    <p class="test">Regular paragraph</p> <br>
    </form>
  </div>
</body>
</html>
```

The output of the code is given below:

Pseudo-Classes (:read-only and :read-write)

In the form the field having type text is enabled and email is disabled

Regular input

Disabled input

Readonly input

Contenteditable paragraph

Regular paragraph

Pseudo-classes (:read-only and :read-write).

PSEUDO-CLASSES (:placeholder-shown)

Example:

```
<!DOCTYPE html>
<html>
<head>
<style>

.demo_container{
  width:600px;
  margin:0 auto;
  justify-content: center;
  align-items: center;
}

input {
  font-size: 1.5rem;
  margin: 10px;
  padding: 10px;
  width: 100%;
}
input:placeholder-shown {
  border: 5px solid red;
}

form {
  display: flex;
  justify-content: center;
  align-items: center;
  flex-direction: column;
}

* {
  box-sizing: border-box;
}

body {
  padding: 1em;
}
</style>
</head>
```

```
<body>
<div class="demo_container">
  <h1> Pseudo-Classes (:placeholder-shown) </h1>
  <form>
    <input type="text" placeholder="Placeholder
text" value=" This is VALUE not PLACEHOLDER ">
    <input type="text" placeholder="This is
PLACEHOLDER not VALUE">

  </form>

  </div>
</body>
</html>
```

The output of the code is given below:

Pseudo-Classes (:placeholder-shown)

This is VALUE not PLACEHOLDER

This is PLACEHOLDER not VALUE

Pseudo-classes (:placeholder-shown).

PSEUDO-CLASSES (:default)

Example:

```
<!DOCTYPE html>
<html>
<head>
<style>

.demo_container{
  width:600px;
  margin:0 auto;
  justify-content: center;
  align-items: center;
}
```

```css
input {
  font-size: 1.5rem;
  margin: 10px;
  padding: 10px;
  width: 100%;
}
input:placeholder-shown {
  border: 5px solid red;
}

form {
  display: flex;
  justify-content: center;
  align-items: center;
  flex-direction: column;
}

* {
  box-sizing: border-box;
}

body {
  padding: 1em;
}
button:default {
    background-color: lime;
    padding: 20px;
    border-radius: 20px;
}
</style>
</head>
<body>
<div class="demo_container">
  <h1> Pseudo-Classes (:default) </h1>
  <form>
    <p> The :default selector selects the default
form element in a group of related elements. </p>
    <button> First Button (It always select the
first element) </button><br>
    <button> Second Button  </button>
  </form>

  </div>
</body>
</html>
```

The output of the code is given below:

Pseudo-Classes (:default)

The :default selector selects the default form element in a group of related elements.

First Button (It always select the first element)

Second Button

Pseudo-classes (:default).

PSEUDO-CLASSES (:checked)

It is associated with input (<input>) elements of type radio and checkbox. The pseudo-class selector matches radio and checkbox input types when checked or toggled to the state. If they are not checked or selected, there is no match.

Example:

```
<!DOCTYPE html>
<html>
<head>
<style>

.demo_container{
  width:600px;
  margin:0 auto;
  justify-content: center;
  align-items: center;
}

input {
  font-size: 1.5rem;
  margin: 10px;
  padding: 10px;
  width: 100%;
}
```

```
input:placeholder-shown {
  border: 5px solid red;
}

form {
  display: flex;
  justify-content: center;
  align-items: center;
  flex-direction: column;
}

* {
  box-sizing: border-box;
}

body {
  padding: 1em;
}
input[type=checkbox] + label {
  color: #ccc;
  font-style: italic;
}
input[type=checkbox]:checked + label {
  color: #f00;
  font-style: normal;
}
</style>
</head>
<body>
<div class="demo_container">
  <h1> Pseudo-Classes (:checked) </h1>
  <form>
    <p> The :checked pseudo-class in CSS selects
elements when they are in the selected state. </p>
    <input type="checkbox" id="title"
name="title">
    <label for="title"> You are using CSS </label>
      </form>

  </div>
</body>
</html>
```

The output of the code is given below:

Pseudo-Classes (:checked)

The :checked pseudo-class in CSS selects elements when they are in the selected state.

☑
You are using CSS

Pseudo-classes (:checked).

PSEUDO-CLASSES (:indeterminate)

The :indeterminate selector is used to select any form elements that are in an indeterminate state, that is, a state that is neither checked nor unchecked.

Example:

```
<!DOCTYPE html>
<html>
<head>
<style>

.demo_container{
  width:600px;
  margin:0 auto;
  justify-content: center;
  align-items: center;
}

form {
  display: flex;
  justify-content: center;
  align-items: center;
  flex-direction: column;
}

* {
  box-sizing: border-box;
}
```

```
body {
  padding: 1em;
}
input:indeterminate + label {
            background: green;
            color: white;
            padding: 4px;
        }

        input:indeterminate {
            box-shadow: 0 0 1px 1px green;
        }
</style>
</head>
<body>
<div class="demo_container">
 <h2>
    CSS :indeterminate selector
</h2>
<div>
<input type="checkbox" id="checkbox">
<label for="checkbox">This is an indeterminate
checkbox.</label>
</div>
<br>
<div>
<input type="radio" id="radio" name="abc">
<label for="radio">This is an indeterminate radio
button.</label>
</div>

<script>
var doc = document.getElementsByTagName("input");

for (var i = 0; i < doc.length; i++) {
    doc[i].indeterminate = true;
}
</script>

   </div>
</body>
</html>
```

The output of the code is given below:

CSS :indeterminate selector

☐ This is an indeterminate checkbox.

◯ This is an indeterminate radio button.

Pseudo-classes (:indeterminate).

PSEUDO-CLASSES (:valid and :invalid)

The :valid selector selects form elements with a value that validates according to the element's settings. The :invalid pseudo-class applies to form fields whose contents do not match the specified type.

Example:

```
<!DOCTYPE html>
<html>
<head>
<style>

.demo_container{
  width:600px;
  margin:0 auto;
  justify-content: center;
  align-items: center;
}

form {
  display: flex;
  justify-content: center;
  align-items: center;
  flex-direction: column;
}

* {
  box-sizing: border-box;
}
```

```
body {
  padding: 1em;
}
input {
  border-radius: 20px;
  width:250px;
   background: #b4f7ab;
   border: 1px solid #06bca7;
   padding: 3px 10px;
   height:50px
}
input:invalid {
   background: #fc9cac;
   border: 1px solid #bc0624;
}

</style>
</head>
<body>
<div class="demo_container">
    <h2> CSS :indeterminate selector </h2>

    <label> Name : </label> <input type="text"
placeholder="Enter text here "> <br>
    <label> Number : </label> <input type="number"
placeholder="Enter number" > <br>
    <label> Email : </label> <input type="email"
placeholder="Enter email here" > <br>

<script>
var doc = document.getElementsByTagName("input");

for (var i = 0; i < doc.length; i++) {
    doc[i].indeterminate = true;
}
</script>

  </div>
</body>
</html>
```

PSEUDO-CLASSES (:in-range and :out-of-range)

The :in-range selects all elements with a value within a specified range in the code and The:out-of-range selector selects all elements with a value outside a specified range in the code.

Example:

```
<!DOCTYPE html>
<html>
<head>
<style>

.demo_container{
  width:600px;
  margin:0 auto;
  justify-content: center;
  align-items: center;
}

form {
  display: flex;
  justify-content: center;
  align-items: center;
  flex-direction: column;
}

* {
  box-sizing: border-box;
}

body {
  padding: 1em;
}
input {
  border-radius: 20px;
  width:250px;
  background: #b4f7ab;
  border: 1px solid #06bca7;
  padding: 3px 10px;
  height:50px
}
```

```
input:in-range {
  border: 5px solid blueviolet ;
}
input:out-of-range {
  border: 5px solid  magenta ;
}

</style>
</head>
<body>
<div class="demo_container">
    <h2> CSS (:in-range and :out-of-range)
selector </h2>
    <p>  Enter the number range between the 5 to
20 to see the styling over the input field</p>
    <input type="number" min="5" max="20"
placeholder="Enter the number">  <span> In range
    ( code of range will run )
    </span> <br> <br>

    <input type="number" min="5" max="20"
placeholder="Enter the number"> <span> Not in
range
    (code of out-of range will run)</span>

  </div>
</body>
</html>
```

The output of the code is given below:

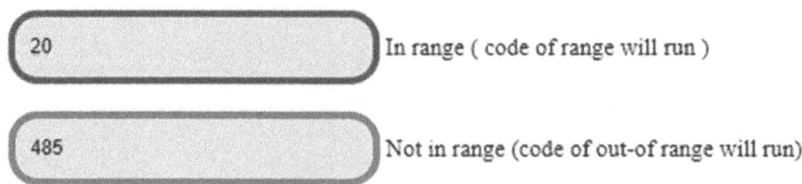

CSS (:in-range and :out-of-range) selector

Enter the number range between the 5 to 20 to see the styling over the input field

Pseudo-classes (:in-range and :out-of-range).

PSEUDO-CLASSES (:required)

Example:

```
<!DOCTYPE html>
<html>
<head>
<style>

.demo_container{
  width:600px;
  margin:0 auto;
  justify-content: center;
  align-items: center;
}

form {
  display: flex;
  justify-content: center;
  align-items: center;
  flex-direction: column;
}

* {
  box-sizing: border-box;
}

body {
  padding: 1em;
}
input {
  border-radius: 20px;
  width:250px;
   border: 1px solid #06bca7;
   padding: 3px 10px;
   height:50px
}

input:required {
  background-color: pink;
}
</style>
</head>
```

```
<body>
<div class="demo_container">
    <h2> CSS Selector (:required)  </h2>
    <input type="number"  placeholder="Enter the
number">
    <br> <br>
    <input type="number"  placeholder="Enter the
number (required ) " required>
  </div>
</body>
</html>
```

PSEUDO-CLASSES (:optional)

Example:

```
<!DOCTYPE html>
<html>
<head>
<style>

.demo_container{
  width:600px;
  margin:0 auto;
  justify-content: center;
  align-items: center;
}

form {
  display: flex;
  justify-content: center;
  align-items: center;
  flex-direction: column;
}

* {
  box-sizing: border-box;
}

body {
  padding: 1em;
}
```

```
input {
  border-radius: 20px;
  width:250px;
   border: 1px solid #06bca7;
   padding: 3px 10px;
   height:50px
}

input:required {
  background-color: lightcoral;
}

input:optional {
  background-color: lightgray;
}
</style>
</head>
<body>
<div class="demo_container">
    <h2> CSS Selector (:optional)  </h2>
    <input type="number"  placeholder="Enter the
number (optional)">
    <br> <br>
    <input type="number"  placeholder="Enter the
number (required)" required>
  </div>
</body>
</html>
```

The output of the code is given below:

CSS Selector (:optional)

Enter the number (optional)

Enter the number (required)

Pseudo-classes (:required and :optional).

PSEUDO-CLASSES (:root)

The :root selector matches the document's root element.

Example:

```
<!DOCTYPE html>
<html>
<head>
<style>

.demo_container{
  width:600px;
  margin:0 auto;
  justify-content: center;
  align-items: center;
}

:root {
  background: #ff0000;
  font-size: 30px;
}
</style>
</head>
<body>
<div class="demo_container">
    <h2> CSS Selector (:root)  </h2>
    <p> It set the background color for the whole
<br>  document and apply all the css as same to
all the elements in HTML </p>

  </div>
</body>
</html>
```

The output of the code is given below:

CSS Selector (:root)

It set the background color for the whole document and apply all the css as same to all the elements in HTML

Pseudo-classes (:root).

PSEUDO-CLASSES (:empty)

It specifies a background color for empty <p> elements.

Example:

```
<!DOCTYPE html>
<html>
<head>
<style>

.demo_container{
  width:600px;
  margin:0 auto;
  justify-content: center;
  align-items: center;
}

p:empty {
  width: 100px;
  height: 20px;
  background: #ff0000;
}

</style>
</head>
```

```
<body>
<div class="demo_container">
    <h2> CSS Selector (:empty)  </h2>
    1. <p></p>
    2. <p> Not Empty </p>
    3. <p> (tags with single space also cannot
considered as empty )</p>
  </div>
</body>
</html>
```

The output of the code is given below:

CSS Selector (:empty)

1.

2.

Not Empty

3.

(tags with single space also cannot considered as empty)

Pseudo-classes (:empty).

PSEUDO-CLASSES (:blank)

Example:

```
<!DOCTYPE html>
<html>
<head>
<style>

.demo_container{
  width:600px;
  margin:0 auto;
  justify-content: center;
  align-items: center;
}
```

```
p {
  min-height: 30px;
  width: 250px;
  background-color: lightblue;
}

p:blank { display: none; }

</style>
</head>
<body>
<div class="demo_container">
    <h2> CSS Selector (:blank)  </h2>
    <div class="blanks">
      <p>This paragraph is not empty or
blank.</p>
      <p><!--this is empty and blank --></p>
      <p>
        <!-- This is not empty, because it has
whitespace. But it is blank.-->
      </p>
      <p>This paragraph is not empty or
blank.</p>
    </div>
  </div>
</body>
</html>
```

The output of the code is given below:

CSS Selector (:blank)

This paragraph is not empty or blank.

This paragraph is not empty or blank.

CSS selector (: blank).

PSEUDO-CLASSES (:nth-child)

Example:

```
<!DOCTYPE html>
<html>
<head>
<style>

.demo_container{
  width:600px;
  margin:0 auto;
  justify-content: center;
  align-items: center;
}

/* It selects the second li element in a list */
li:nth-child(2) {
  background: light green;
}

/* It selects every third element among any group
of siblings */
:nth-child(3) {
  background: yellow;
}

</style>
</head>
<body>
<div class="demo_container">
  <h2>The :nth-child(n) selector matches
every element that is the nth child of its
parent. </h2>
  <div>
    <p>This is some text.</p>
  </div>

  <div>
    <p>This is some text.</p>
  </div>
```

```
<div>
  <p>This is some text.</p>
</div>

<ul>
  <li>First list item</li>
  <li>Second list item</li>
  <li>Third list item</li>
  <li>Fourth list item</li>
  <li>Fifth list item</li>
</ul>

</div>
</body>
</html>
```

The output of the code is given below:

The :nth-child(n) selector matches every element that is the nth child of its parent.

- First list item
- Second list item
- Third list item
- Fourth list item
- Fifth list item

Pseudo-classes (:nth-child).

PSEUDO-CLASSES (:nth-last-child)

Example:

```
<!DOCTYPE html>
<html>
<head>
<style>

.demo_container{
  width:600px;
  margin:0 auto;
  justify-content: center;
  align-items: center;
}
```

```
/* Selects the second li element in a list */
li:nth-child(2) {
  background: red;
}

/* Selects every third element among any group of
siblings */
li:nth-child(3) {
  background: yellow;
}

</style>
</head>
<body>
<div class="demo_container">
  <h2>The :nth-last-child(n) selector matches
last element that is the nth child of its parent.
</h2>
  <ul>
    <li>First list item</li>
    <li>Second list item</li>
    <li>Third list item</li>
    <li>Fourth list item</li>
    <li>Fifth list item</li>
  </ul>

  </div>
</body>
</html>
```

The output of the code is given below:

The :nth-last-child(n) selector matches last element that is the nth child of its parent.

- First list item
- Second list item
- Third list item
- Fourth list item
- Fifth list item

Pseudo-classes (:nth-child).

PSEUDO-CLASSES (:first-child)

Example:

```
<!DOCTYPE html>
<html>
<head>
<style>

.demo_container{
  width:600px;
  margin:0 auto;
  justify-content: center;
  align-items: center;
}

/* Selects the second li element in a list */
li:first-child{
  background: red;
}


</style>
</head>
<body>
<div class="demo_container">
  <h2>The :first-child selector matches the
first element that is the nth child of its
parent. </h2>
  <ul>
    <li>First list item</li>
    <li>Second list item</li>
    <li>Third list item</li>
    <li>Fourth list item</li>
    <li>Fifth list item</li>
  </ul>

  </div>
</body>
</html>
```

The output of the code is given below:

The :first-child selector matches first element that is the nth child of its parent.

- First list item
- Second list item
- Third list item
- Fourth list item
- Fifth list item

Pseudo-classes (:first-child).

PSEUDO-CLASSES (:only-child)

Example:

```
<!DOCTYPE html>
<html>
<head>
<style>

.demo_container{
  width:600px;
  margin:0 auto;
  justify-content: center;
  align-items: center;
}

/* Selects the second li element in a list */
li:only-child{
  background: red;
}


</style>
</head>
<body>
<div class="demo_container">
  <h2>The :only-child selectors selects an
element, only child of its parent. </h2>
```

```
    <h3>List 1</h3>
    <ul>
      <li>First list item</li>
      <li>Second list item</li>
      <li>Third list item</li>
      <li>Fourth list item</li>
      <li>Fifth list item</li>
    </ul>

    <h3>List 2</h3>
    <ul>
      <li>First list item</li>
    </ul>

  </div>
</body>
</html>
```

The output of the code is given below:

The :only-child selectors selects an element, only child of its parent.

List 1

- First list item
- Second list item
- Third list item
- Fourth list item
- Fifth list item

List 2

- First list item

Pseudo-classes (:only-child).

PSEUDO-CLASSES (:nth-of-type(odd)

Example:

```
<!DOCTYPE html>
<html>
<head>
<style>
```

```css
.demo_container{
  width:600px;
  margin:0 auto;
  justify-content: center;
  align-items: center;
}

/* Selects the second li element in a list */
li:nth-of-type(odd){
  width:200px;
  background: red;
}

li:nth-of-type(even){
  width:200px;
  background: green;
}

</style>
</head>
<body>
<div class="demo_container">
  <h2>The :only-child selectors selects an
element, only child of its parent. </h2>

  <h3>List 1</h3>
  <ul>
    <li>First list item</li>
    <li>Second list item</li>
    <li>Third list item</li>
    <li>Fourth list item</li>
    <li>Fifth list item</li>
  </ul>

  <h3>List 2</h3>
  <ul>
    <li>First list item</li>
  </ul>

  </div>
</body>
</html>
```

PSEUDO-CLASSES (:nth-of-type(odd or even)

The: nth-of-type(n) selector matches every element, that is, the nth child, of the same type (tag name), of its parent.

Example:

```
<!DOCTYPE html>
<html>
<head>
<style>

.demo_container{
  width:600px;
  margin:0 auto;
  justify-content: center;
  align-items: center;
}

/* Selects the second li element in a list */
li:nth-of-type(odd){
  width:200px;
  background: red;
}

li:nth-of-type(even){
  width:200px;
  background: green;
}

</style>
</head>
<body>
<div class="demo_container">
  <h2>The (:nth-of-type(odd), nth-of-type(even))
selectors selects an element, only child of its
parent. </h2>

  <h3>List 1</h3>
  <ul>
    <li>First list item</li>
    <li>Second list item</li>
    <li>Third list item</li>
```

```
   <li>Fourth list item</li>
   <li>Fifth list item</li>
</ul>

<h3>List 2</h3>
<ul>
   <li>First list item</li>
</ul>

</div>
</body>
</html>
```

The output of the ode is given below:

The (:nth-of-type(odd), nth-of-type(even)) selectors selects an element, only child of its parent.

List 1

- First list item
- Second list item
- Third list item
- Fourth list item
- Fifth list item

List 2

- First list item

Pseudo-classes (:nth-of-type(odd or even)).

PSEUDO-CLASSES (:first-of-type, nth-of-type, nth-last-of-type(3))

Example:

```
<!DOCTYPE html>
<html>
<head>
<style>
```

```css
.demo_container{
  width:600px;
  margin:0 auto;
  justify-content: center;
  align-items: center;
}

/* Selects the second li element in a list */
li:first-of-type{
  width:200px;
  background: red;
}

li:nth-of-type(even){
  width:200px;
  background: green;
}

li:nth-last-of-type(3){
  width:200px;
  background: yellow;
}

</style>
</head>
<body>
<div class="demo_container">
  <h2>The (:first-of-type, nth-of-type, nth-last-
of-type(3)) selectors selects an element, only
child of its parent. </h2>

  <h3>List 1</h3>
  <ul>
    <li>First list item</li>
    <li>Second list item</li>
    <li>Third list item</li>
    <li>Fourth list item</li>
    <li>Fifth list item</li>
  </ul>
```

```
<h3>List 2</h3>
<ul>
  <li>First list item</li>
</ul>

</div>
</body>
</html>
```

The output of the code is given below:

The (:first-of-type, nth-of-type, nth-last-of-type(3)) selectors selects an element, only child of its parent.

List 1

- First list item
- Second list item
- Third list item
- Fourth list item
- Fifth list item

List 2

- First list item

Pseudo-classes (Pseudo-classes (:first-of-type, nth-of-type, nth-last-of-type(3)).

COMBINATORS SELECTORS

COMBINATORS SELECTORS (E F)

It select and style all <h1> elements AND all <h4> elements.

Example:

```
<!DOCTYPE html>
<html>
<head>
<style>

.demo_container{
  width:600px;
  margin:0 auto;
```

```
    justify-content: center;
    align-items: center;
}

h1{
    background-color: pink;
}
h4{
    background-color: lightblue;
}

</style>
</head>
<body>
<div class="demo_container">
  <h2>It select and style all h1 elements and all
h4 elements. </h2>


  <h1>List 1</h1>
  <ul>
    <li>First list item</li>
    <li>Second list item</li>
    <li>Third list item</li>
    <li>Fourth list item</li>
    <li>Fifth list item</li>
  </ul>

  <h4>List 2</h4>
  <ul>
    <li>First list item</li>
  </ul>


  <h1> This is H1 Heading </h1>

  <h4> This is H4 Heading </h4>

  </div>
</body>
</html>
```

It is also called as a descendant selector (space).

The output of the code is given below:

It select and style all h1 elements and all h4 elements.

List 1

- First list item
- Second list item
- Third list item
- Fourth list item
- Fifth list item

List 2

- First list item

This is H1 Heading

This is H4 Heading

Combinators selectors (E F).

COMBINATORS SELECTORS (E > F)

It is a child selector (>). It selects all elements that are the children of a specified element.

Example:

```
<!DOCTYPE html>
<html>
<head>
<style>

.demo_container{
  width:600px;
  margin:0 auto;
  justify-content: center;
  align-items: center;
}
```

```
ul > li {
  background-color: pink;
  border:2px solid darkgreen;
}

</style>
</head>
<body>
<div class="demo_container">
  <h2>It selects all elements that are the
children of a specified element </h2>


  <h1>List 1</h1>
  <ul>
    <li>First list item</li>
    <li>Second list item</li>
    <li>Third list item</li>
    <li>Fourth list item</li>
    <li>Fifth list item</li>
  </ul>

  <h4>List 2</h4>
  <ul>
    <li>First list item</li>
  </ul>


  <h1> This is H1 Heading </h1>

  <h4> This is H4 Heading </h4>

  </div>
</body>
</html>
```

The output of the code is given below:

It selects all elements that are the children of a specified element

List 1

- First list item
- Second list item
- Third list item
- Fourth list item
- Fifth list item

List 2

- First list item

This is H1 Heading

This is H4 Heading

Combinators selectors (E > F).

COMBINATORS SELECTORS (E + F)

The adjacent sibling selector selects an element that is directly after another specific element. It is also called adjacent sibling selector (+).

Example:

```
<!DOCTYPE html>
<html>
<head>
<style>

.demo_container{
  width:600px;
  margin:0 auto;
  justify-content: center;
  align-items: center;
}
```

```
h1 + h4 {
  background-color: pink;
  border:2px solid darkgreen;
}

</style>
</head>
<body>
<div class="demo_container">
  <h2>The + selector is used to select an
element that is directly after another specific
element.</h2>


  <h1>List 1</h1>
  <ul>
    <li>First list item</li>
    <li>Second list item</li>
  </ul>

  <h4>List 2</h4>
  <ul>
    <li>First list item</li>
  </ul>


  <h1> This is H1 Heading </h1>

  <h4> This is H4 Heading </h4>

  <h1> This is H1 Heading </h1>

  <h4> This is H4 Heading </h4>

  </div>
</body>
</html>
```

The output of the code is given below:

The + selector is used to select an element that is directly after another specific element.

List 1

- First list item
- Second list item

List 2

- First list item

This is H1 Heading

This is H4 Heading

This is H1 Heading

This is H4 Heading

Combinators selectors (E + F).

GENERAL SIBLING SELECTOR (~)

The general sibling selects all elements that are the next siblings of a specified element.

Example:

```
<!DOCTYPE html>
<html>
<head>
<style>

.demo_container{
    width:600px;
    margin:0 auto;
    justify-content: center;
    align-items: center;
}
```

```
h1 ~ h4 {
  background-color: yellow;
  border:2px solid darkgreen;
}

</style>
</head>
<body>
<div class="demo_container">
  <h2>The  (~) selects all elements that are next
siblings of a specified element.</h2>


  <h1>List 1</h1>
  <ul>
    <li>First list item</li>
    <li>Second list item</li>
  </ul>

  <h4>List 2</h4>
  <ul>
    <li>First list item</li>
  </ul>


  <h1> This is H1 Heading </h1>

  <h4> This is H4 Heading </h4>

  <h1> This is H1 Heading </h1>

  <h4> This is H4 Heading </h4>

  </div>
</body>
</html>
```

The output of the code is given below:

The (~) selects all elements that are next siblings of a specified element.

List 1

- First list item
- Second list item

List 2

- First list item

This is H1 Heading

This is H4 Heading

This is H1 Heading

This is H4 Heading

Combinators selectors (E ~ F).

GRID-STRUCTURAL SELECTORS (:nth-col())

The grid-structural pseudo-class is made for structural grids like tables. The :nth-col() pseudo-class represents a cell belonging to the column in a grid structure that has nth-columns before it, while :nth-last-col() will count from after that column.

Syntax:

```
:nth-col(An+B) {
    /* declarations */
}

:nth-last-col(An+B) {
    /* declarations */
}
```

Example:

```html
<!DOCTYPE html>
<html>
<head>
<style>

.demo_container{
  width:600px;
  margin:0 auto;
  justify-content: center;
  align-items: center;
}


:nth-col(2n+1) {
    background-color: gray;
}

:nth-last-col(3n+1) {
    background-color: green;
}
</style>
</head>
<body>
<div class="demo_container">
  <table>
    <h1>Grid-Structural Selectors    </h1>
    <col />
    <col class="highlight" />
    <col />
    <tr>
        <td>A</td>
        <td>B</td>
        <td>C</td>
    </tr>
    <tr>
        <td>C</td>
        <td>D</td>
        <td>E</td>
    </tr>
```

```
    <tr>
        <td>F</td>
        <td>G</td>
        <td>H</td>
    </tr>
    <tr>
        <td>I</td>
        <td>J</td>
        <td>K</td>
    </tr>
    <tr>
        <td>L</td>
        <td>M</td>
        <td>N</td>
    </tr>
    <tr>
        <td>O</td>
        <td>P</td>
        <td>Q</td>
    </tr>
</table>
<p> Now, It doesn't work on any of the browser</p>
    </div>
</body>
</html>
```

The output of the code is given below:

Grid-Structural Selectors (:nth-col())

```
A B C
C D E
F G H
I J K
L M N
O P Q
```

Now, It doesn't work on any of the browser

Grid-structural selectors (:nth-col()).

CHAPTER SUMMARY

In this chapter, we covered all the valuable properties and mainly used properties of CSS with descriptions. By reading this, the beginner-level user can have good knowledge. The next chapter is about CSS built-in functions.

CSS Functions

INTRODUCTION

CSS functions are used for various CSS properties. For example, you can use the function RGB() to provide a color value (such as color: RGB(105, 0, 215)), or the attr() function to get the value of an HTML attribute.

Various functions are used in CSS transforms. For example, the rotate() function can be used to rotate an element, the scale() function can be used to change the size of an element, and the translate() function can be used to move an element.

There are also some functions, such as circle() to clip an element to a circle or create a circle for text to flow around, the calc() function can provide a calculated value for a property.

DOI: 10.1201/9781003358060-4

BASICS OF CSS FUNCTIONS

attr()

The attr() function retrieves the value of the selected element and can be used in the stylesheet. It can be used on pseudo elements, in which case the value of the attribute on the pseudo element's originating element is returned.

Example:

```
<!DOCTYPE html>
<html>
<head>
<style>

* {
  box-sizing: border-box;
}

body {
  padding: 1em;
}
.demo_container{
  width:600px;
  margin:0 auto;
  justify-content: center;
  align-items: center;
}

  abbr[title]:after {
    content: " (" attr(title) ")";
  }


</style>
</head>
<body>
<div class="demo_container">
  <h1> CSS Function ( :attr() ) </h1>
  <p> This attr() function to return and display
the title attribute from an abbreviation given
below </p>
```

```
<abbr title="World Health Organization">
  WHO
</abbr>

</div>
</body>
</html>
```

The output of the above code is given below:

CSS Function (:attr())

The attr() function to return and display the title attribute from an abbreviation given below

WHO (World Health Organization)

CSS function (:attr()).

Usage of attr()
Another example of a URL of a hyperlink.

```
<!DOCTYPE html>
<html>
<head>
<style>

* {
  box-sizing: border-box;
}

body {
  padding: 1em;
}
.demo_container{
  width:600px;
  margin:0 auto;
  justify-content: center;
  align-items: center;
}
```

```
a[href]::after {
    content: " [ " attr(href) " ] ";
    font-style: italic;
  }

</style>
</head>
<body>
<div class="demo_container">
  <h1> CSS Function ( :attr() ) </h1>
  <p> Here the attr() function can add any other text
around the value
    (such as enclose it in brackets, adding spaces,
and so on). <br>
  </p>
  <a href="https://www.google.com">
    Google Page
  </a>

  </div>
</body>
</html>
```

The output of the code is given below:

CSS Function (:attr())

Here in the attr() function you can add any other text around the value (such as enclosing it in brackets, adding spaces, and so on).

Google Page [https://www.google.com]

CSS function (:attr()).

Here is the example in which you will get to know how to append the text using attr()CSS.

```
<!DOCTYPE html>
<html>
<head>
<style>

* {
  box-sizing: border-box;
}
```

```
body {
  padding: 1em;
}
.demo_container{
  width:600px;
  margin:0 auto;
  justify-content: center;
  align-items: center;
}
li:after {
    content: " (" attr(data-sitcom) ")";
  }

</style>
</head>
<body>
<div class="demo_container">
  <h3> In the given list the content inside the attr()
function will be appended to the list data </h3>
  <ul>
    <li data-sitcom="Extra Content 1">
      A
    </li>
    <li data-sitcom="Extra Content 2">
      B
    </li>
    <li>
      C ( There is not data-sitcom of C ) ->
    </li>
    <li data-sitcom="Extra Content 3">
      D
    </li>
    <li data-sitcom="Extra Content 4">
      E
    </li>
    <li data-sitcom="Extra Content 5">
      F
    </li>
  </ul>

  </div>
</body>
</html>
```

The output of the code is given below:

In the given list the content inside the attr() function will be appended to the list data

- A (Extra Content 1)
- B (Extra Content 2)
- C (There is not data-sitcom of C) -> ()
- D (Extra Content 3)
- E (Extra Content 4)
- F (Extra Content 5)

CSS function (:attr()).

blur()

The blur() is used with the filter to apply a blurred effect to an image. The syntax of the blur() function is as follows:

```
blur() = blur( <length> )
```

Example:

```html
<!DOCTYPE html>
<html>
<head>
<style>

* {
  padding:0;
  margin:0;
  box-sizing: border-box;
}

.demo_container{
  padding:20px;
  width:600px;
  margin:0 auto;
  justify-content: center;
  align-items: center;
}

img{
  width:400px;
  height:200px;
}
```

```
.filtered {
    filter: blur(3px);
  }
</style>
</head>
<body>
<div class="demo_container">
  <h3> The blur() function will blur the given
image. </h3>
  <p>Normal Image</p>
  <img src="/images-1.jpg" alt="Sample image">
  <p>Blur Image</p>
  <img class="filtered" src="/images-1.jpg"
alt="Sample image">
  </div>
</body>
</html>
```

The output of the code is given below:

The blur() function will blur the given image.
Normal Image

Blur Image

CSS function (:blur()).

brightness()

The brightness() function is used to adjust an image's brightness.

The syntax of the brightness() function is as follows:

```
brightness() = brightness( [ <number> | <percentage> ] )
```

Explanation: The brightness() function accepts a number or also percentage as its argument. It determines the brightness level of the image. A value of 0% in the image will be completely black. A value of 100% in an image will be unchanged. The value above 100% produces a brighter image. A number value of 0.2 has the same effect as the percentage value of 20%. The negative values are not allowed. It is used with the filter property to adjust the brightness of an image. The function applies a linear multiplier to the input image, making it appear more or less bright. It requires an argument to be passed to it. It determines the brightness level that is applied to the image. The argument can be either a % value or a number.

Example:

```
<!DOCTYPE html>
<html>
<head>
<style>

* {
  padding:0;
  margin:0;
  box-sizing: border-box;
}

.demo_container{
  padding:20px;
  width:800px;
  margin:0 auto;
  justify-content: center;
  align-items: center;
}
```

```
img{
  width:100%;
  height:250px;
}
.row{
  display: flex;
}
.col{

width:100%
}
.image-1 {
  filter: brightness(50%);
  }
  .image-2 {
  filter: brightness(150%);
  }
  .image-3 {
  filter: brightness(2.5);
  }
  p{
    padding-top: 20px;
    font-size:20px
  }
</style>
</head>
<body>
<div class="demo_container">
  <h2> The blur() function will blur the given
image. </h2>
  <div class="row">
    <div class="col">
      <p> Normal Image </p>
      <img src="/images-1.jpg" alt="Sample image">
      <p> 50% Brightness Image </p>
      <img class="image-1" src="/images-1.jpg"
alt="Sample image">
    </div>
    <div class="col">
      <p> 150% Brightness Image </p>
    <img class="image-2" src="/images-1.jpg"
alt="Sample image">
```

```
    <p> 50% Brightness Image </p>
    <img class="image-3" src="/images-1.jpg"
alt="Sample image">
    </div>
  </div>
  </div>
</body>
</html>
```

The output of the above code is given below:

The blur() function will blur the given image.

Normal Image   150% Brightness Image

50% Brightness Image   50% Brightness Image

CSS function (:blur()).

calc()

The CSS calc() function allows the use of calculations within CSS property values. The calc() can be used in place of other unit types when setting widths, heights, angles, frequencies, etc. The exact value that the browser uses will be a result of the calculation performed by the calc() function.

Example:

```html
<!DOCTYPE html>
<html>
<head>
<style>

* {
  padding:0;
  margin:0;
  box-sizing: border-box;
}

.demo_container{
  width:800px;
  margin:0 auto;
  justify-content: center;
  align-items: center;
}

  nav {
    width: 180px;
    float: left;
    background: gold;
  }
  article {
    width: calc(90% - 100px);
    float: right;
    background: orange;
  }
  article, nav {
    color: white;
    padding: 30px;
    box-sizing: border-box;
  }
</style>
<div class="demo_container">
  <h1> The calc() function allows to use
calculations within CSS property values. </h1>
<br>
  <nav>
    <h1> Width:100px </h1>
  </nav>
```

```
<article>
  <h1> Width: calc(100% - 100px);</h1>
  </article>
</div>

</body>
</html>
```

The output of the above code is given below:

The calc() function allows to use calculations within CSS property values.

| Width:100px | Width: calc(100% – 100px); |

CSS function (:calc ()).

circle()

The circle() is a basic shape value that is the part of the CSS shapes module. Basic shapes such as circle() can be used as a value for properties such as shape-outside to control the flow of content (code) around the element and clip-path to clip the element's contents to the basic shape. It means that you have text flowing around the element in the shape of a circle, has an image clipped to the shape of a circle.

The syntax of the circle() function is as follows:

```
circle() = circle( [<shape-radius>]? [at <position>]? )
```

How to Position the Circle

You can specify a position for the circle by various radius argument with a valid CSS position. It specifies the circle's center. It just separates the radius and the position like this:

```
shape-outside: circle(100px at 10px 150px);
```

Example:

```
<!DOCTYPE html>
<html>
<head>
<style>
```

```css
* {
  padding:0;
  margin:0;
  box-sizing: border-box;
}

.demo_container{
  width:800px;
  margin:0 auto;
  justify-content: center;
  align-items: center;
}

.section-1 {
  float: left;
  width: 200px;
  height: 150px;
  shape-outside: circle();
}
.section-2 {
  float: left;
  width: 200px;
  height: 150px;
  shape-outside: circle(100px at 0px 150px);
}
</style>
<div class="demo_container">
  <h1> The circle() is a CSS shape value that's
part of the CSS Shapes module circle(with no
background-color). </h1> <br>
  <section class="section-1"> </section>
  <p>Ornare quam viver raorci sagittis eu volutpat
odio. Viverra adipiscing at inellus integer
feugiat scelerisque.
    Adipiscing biben dum est ultricies integer quis
auctor. Massa tincidunt dui utornare lectus sit amet.
    Pellentesque eliteget gravida cum sociis
natoque penatibus et. Sed vulputate odio ut enim
blandit volutpat maecenas volutpat.
    Purus viverra accumsan in nisl nisi. Dignissim
enim sit amet venenatis urnacu rsus eget. Ornare
arcuo dio ut sem nullap haretra diam sit.
    Vitae justo eget magna fermentum iaculis.</p>
</div> <br> <br> <br>
```

```
<div class="demo_container">
  <h1> The circle() is a CSS shape value that's
part of the CSS Shapes module (with background-
color). </h1> <br>
  <section class="section-2"></section>
  <p>Ornare quam viverra orci sagittis euvo lutpat
odio. Viverra adipiscing at int ellus integer
feugiat scelerisque.
    Adipiscing bibendum est ultricies integ erquis
auctor. Massa tincidunt du iut ornare lectus sit
amet.
    Pellentesque eli teget gravi dacum sociis
natoque penatibus et. Sed vulputate odio utenim
blandit volutpat maecenas volutpat.
    Purus viverraac cumsan in nisl nisi.
Dignissim eni msit amet venenatis urna cursus
eget. Orna rearcu odio ut sem nulla pharetra diam
sit.
    Vitae justoeget magna fermentum iaculis.</p>
</div>

</body>
</html>
```

The output of the code is given below:

The circle() is a CSS shape value that's part of the CSS Shapes module circle(with no backgroud-color).

Ornare quam viverra orci sagittis eu volutpat odio. Viverra adipiscing at in tellus integer feugiat scelerisque. Adipiscing bibendum est ultricies integer quis auctor. Massa tincidunt dui ut ornare lectus sit amet. Pellentesque elit eget gravida cum sociis natoque penatibus et. Sed vulputate odio ut enim blandit volutpat maecenas volutpat. Purus viverra accumsan in nisl nisi. Dignissim enim sit amet venenatis urna cursus eget. Ornare arcu odio ut sem nulla pharetra diam sit. Vitae justo eget magna fermentum iaculis.

The circle() is a CSS shape value that's part of the CSS Shapes module (with backgroud-color).

Ornare quam viverra orci sagittis eu volutpat odio. Viverra adipiscing at in tellus integer feugiat scelerisque. Adipiscing bibendum est ultricies integer quis auctor. Massa tincidunt dui ut ornare lectus sit amet. Pellentesque elit eget gravida cum sociis natoque penatibus et. Sed vulputate odio ut enim blandit volutpat maecenas volutpat. Purus viverra accumsan in nisl nisi. Dignissim enim sit amet venenatis urna cursus eget. Ornare arcu odio ut sem nulla pharetra diam sit. Vitae justo eget magna fermentum iaculis.

CSS function (:circle()).

Another example:

```
<!DOCTYPE html>
<html>
<head>
<style>

* {
  padding:0;
  margin:0;
  box-sizing: border-box;
}

.demo_container{
  width:800px;
  margin:0 auto;
  justify-content: center;
  align-items: center;
}

.section-1 {
  float: left;
  width: 200px;
  height: 150px;
  shape-outside: circle();
}
.section-2 {
  float: left;
  width: 200px;
  height: 150px;
  /* shape-outside: circle(100px at 0px 150px); */
  background: brown;
  clip-path: circle();

}
</style>
<div class="demo_container">
  <h1> The circle() is a CSS shape value that's
part of the CSS Shapes module clip-path:
circle()  </h1> <br>
  <section class="section-2"></section>
  <p>Ornare quam viver raorci sagittis eu volutpat
odio. Viverra adipiscing at in tell usinteger
feugiat scelerisque.
```

```
        Adipiscing bibendum esult ricies integer
quisauctor. Massa tincidunt dui ut ornare sit amet.
        Pellentesque elite get gravida cum sociis
natoque et. Sed vulputate odio ut enim blandit
volutpat maecenas volutpat.
        Purus viverra accumsan innisl nisi. Dignissim
enim sitamet venenatis urna cursus eget. Ornare
arc uodio ut sem nulla pharetra diam sit.
        <p>Faucibus ornare suspendisse sed nisi lacus
viverra tellus.
        Rhoncus est pellentesque elit ullamcorper
dignissim cras. Sodales ut etiam sit ame tisl
purus in mollis nunc.
            Hendrerit gravi darutrum quisque nontel
lus orci ac auctor. Enim ut sem viverra eget. Eu
nisl nunc mi ipsum faucibus.
            Enim facilisis gravida neque convallis a
cras semper auctor neque.
        Sit amet mauris commodo quis imperdiet
tincidunt nunc.</p>
        Vitae justo eget magna fermentum iaculis.</p>
</div>

</body>
</html>
```

The output of the code is given below:

The circle() is a CSS shape value that's part of the CSS Shapes module circle(with no backgroud-color).

Ornare quam viverra orci sagittis eu volutpat odio. Viverra adipiscing at in tellus integer feugiat scelerisque. Adipiscing bibendum est ultricies integer quis auctor. Massa tincidunt dui ut ornare lectus sit amet. Pellentesque elit eget gravida cum sociis natoque penatibus et. Sed vulputate odio ut enim blandit volutpat maecenas volutpat. Purus viverra accumsan in nisl nisi. Dignissim enim sit amet venenatis urna cursus eget. Ornare arcu odio ut sem nulla pharetra diam sit. Vitae justo eget magna fermentum iaculis.

The circle() is a CSS shape value that's part of the CSS Shapes module (with backgroud-color).

Ornare quam viverra orci sagittis eu volutpat odio. Viverra adipiscing at in tellus integer feugiat scelerisque. Adipiscing bibendum est ultricies integer quis auctor. Massa tincidunt dui ut ornare lectus sit amet. Pellentesque elit eget gravida cum sociis natoque penatibus et. Sed vulputate odio ut enim blandit volutpat maecenas volutpat. Purus viverra accumsan in nisl nisi. Dignissim enim sit amet venenatis urna cursus eget. Ornare arcu odio ut sem nulla pharetra diam sit. Vitae justo eget magna fermentum iaculis.

CSS function (:circle()).

Another example:

```html
<!DOCTYPE html>
<html>
<head>
<style>

* {
  padding:0;
  margin:0;
  box-sizing: border-box;
}

.demo_container{
  width:800px;
  margin:0 auto;
  justify-content: center;
  align-items: center;
}

.section-1 {
  float: left;
  width: 200px;
  height: 150px;
  shape-outside: circle();
}
.section-2 {
  float: left;
  width: 200px;
  height: 150px;
  background: brown;
  clip-path: circle();

}
</style>
<div class="demo_container">
  <h1> The circle() is a CSS shape value that's
part of the CSS Shapes module clip-path:
circle()  </h1> <br>
  <section class="section-2"></section>
  <p>Ornare quam viverra orci sagit tiseu volutpat
odio. Viverra adipiscing atin tellus integer
feugiat scelerisque.
```

```
        Adipiscing bibendum estul tricies integer quis
auctor. Massa tincidunt dui utornare lectus sit
amet.
        Pellentesque elit egetgravida cum sociis
natoque penatibus et. Sed vulputate odio ut enim
blandit volutpat maecenas volutpat.
        Purus viverra accumsan in nisl nisi.
Dignissim enim sitamet venenatis urna cursus
egt. Ornare arcu odio uts em nulla pharetra diam
sit.
        <p>Faucibus ornare suspendisse sednisi lacus
sed viverra tellus.
        Rhoncus est pellentesque elit ullamcorper
dignissim cras. Sodales ut etiam sit amet nisl
puruin mollis nunc.
        Hendrerit gravida rutrum quisque non
tellus orciac auctor. Enim ut sem viverra aliquet
eget. Eu nisl nunc mi ipsum faucibus.
        Enim facilisis gravida neque convallis
acras semper auctor neque.
        Sit amet mauris commodo quis imperdiet
massa tinci dunt nunc.</p>
        Vitae justo eget magna fermentum iaculis.</p>
</div>

</body>
</html>
```

The output of the code is given below:

The circle() is a CSS shape value that's part of the CSS Shapes module clip-path: circle()

Ornare quam viverra orci sagittis eu volutpat odio. Viverra adipiscing at in tellus integer feugiat scelerisque. Adipiscing bibendum est ultricies integer quis auctor. Massa tincidunt dui ut ornare lectus sit amet. Pellentesque elit eget gravida cum sociis natoque penatibus et. Sed vulputate odio ut enim blandit volutpat maecenas volutpat. Purus viverra accumsan in nisl nisi. Dignissim enim sit amet venenatis urna cursus eget. Ornare arcu odio ut sem nulla pharetra diam sit.

Faucibus ornare suspendisse sed nisi lacus sed viverra tellus. Rhoncus est pellentesque elit ullamcorper dignissim cras. Sodales ut etiam sit amet nisl purus in mollis nunc. Hendrerit gravida rutrum quisque non tellus orci ac auctor. Enim ut sem viverra aliquet eget. Eu nisl nunc mi ipsum faucibus. Enim facilisis gravida neque convallis a cras semper auctor neque. Sit amet mauris commodo quis imperdiet massa tincidunt nunc.

Vitae justo eget magna fermentum iaculis.

CSS function (:circle()).

contrast()

The CSS contrast() is used with the filter property to adjust the contrast of an image. The contrast() requires an argument to be passed to it. It determines the contrast level that's applied to the image. The argument can be either a % value or a number.

The syntax of the contrast() function is as follows:

```
contrast() = contrast( [ <number> | <percentage> ] )
```

Example:

```
<!DOCTYPE html>
<html>
<head>
<style>

* {
  padding:0;
  margin:0;
  box-sizing: border-box;
}

.demo_container{
  padding:20px;
  width:800px;
  margin:0 auto;
  justify-content: center;
  align-items: center;
}

img{
  width:100%;
  height:250px;
}
.row{
  display: flex;
}
.col{

width:100%
}
```

```
.image-1 {
  filter: contrast(140%);
  }
  .image-2 {
    filter: contrast(40%);
}
  .image-3 {
    filter: contrast(2.5);
}
  p{
    font-size:20px
  }
</style>
</head>
<body>
<div class="demo_container">
  <h3> The contrast() functionused with the filter
property to adjust the contrast on given image. </h3>
  <div class="row">
    <div class="col">
      <p> Normal Image </p>
      <img src="/images-1.jpg" alt="Sample image">
      <p> 140% Contrast Image </p>
      <img class="image-1" src="/images-1.jpg"
alt="Sample image">
    </div>
    <div class="col">
      <p> 40% Contrast Image </p>
    <img class="image-2" src="/images-1.jpg"
alt="Sample image">
    <p> 2.5 Contrast Image </p>
    <img class="image-3" src="/images-1.jpg"
alt="Sample image">
    </div>
  </div>
  </div>


</body>
</html>
```

The output of the code is given below:

The contrast() functionused with the filter property to adjust the contrast on given image.

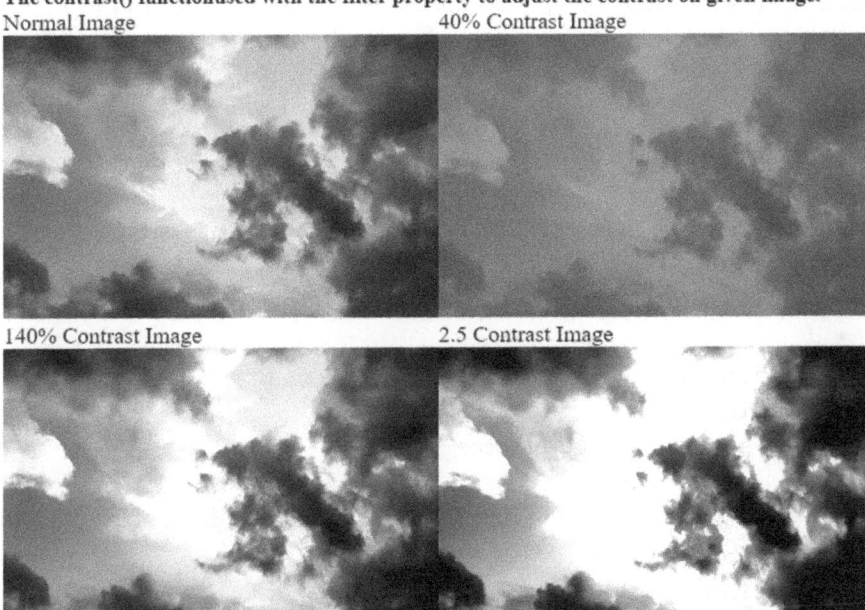

CSS function (:contrast()).

counter()

The counter() allows you to display the counter that was generated by the element. Each element has a collection of multiple counters that are inherited from the document tree much like property values. You can create and manipulate counters using the counter increment, counter reset, and counter set properties. They have no visible effect by themselves but can be used with the counter() and counters() functions to display counter values in a specified format.

For example, you can use the counter() to output an element's counter in Upper Roman, Decimal, Georgian, etc. You can also specify Unicode code points to specify special symbols or icons to use as mark representations.

Syntax: The counter() function has two forms such as: counter(name) or counter(name, style). The text is the value of the innermost counter of the given name in scope at the pseudo-element; it is formatted in the indicated style (decimal by default).

The syntax of the counter() function is as follows:

```
counter( <ident> [, [ <counter-style> | none ] ]? )
```

Example:

```
<!DOCTYPE html>
<html>
<head>
<style>

* {
  padding:0;
  margin:0;
  box-sizing: border-box;
}

.demo_container{
  padding:20px;
  width:800px;
  margin:0 auto;
  justify-content: center;
  align-items: center;
}

section {
    line-height: 0.8em;
    counter-reset: myCounter;
  }
  h1 {
    font-size: 18px;
    }
  p:before {
    content: counter(myCounter) ". ";
    counter-increment: myCounter;
  }

</style>
</head>
```

```
<body>
<div class="demo_container">
  <h3> The counter() enables to display the counter
that has been generated by an element. </h3>
  <section>
  The list of programming is given below:
  <p>HTML</p>
  <p>PHP</p>
  <p>JavaScript</p>
  <p>Ruby</p>

</section>
  </div>
</body>
</html>
```

The output of the code is given below:

The counter() enables to display the counter that has been generated by an element.
The list of programming is given below:
1. HTML
2. PHP
3. JavaScript
4. Ruby

CSS function (:counter()).

counters()

The counters() enables you to display nested counters that have been generated by an element and its parent/s.

Every element has a collection of more counters, which are inherited from the document tree in a way similar to property values. You can create and manipulate counters with the counter-increment, counter-reset, counter-set properties. It has no visible effect by itself, but it can be used with the counters() and counter() functions, which allow the values of counters to be displayed, in the format you specify.

The counters() have two forms: counters(name, string) or counters(such as name, string, style). The text is the value of the counters with the given name in scope at pseudo-element, from outermost to innermost by the specified string. The counters are rendered in the indicated style (decimal by default).

The syntax of the counters() function is as follows:

```
counters( <ident>, <string> [, [ <counter-style> |
none ] ]? )
```

Example:

```
<!DOCTYPE html>
<html>
<head>
<style>

* {
  padding:0;
  margin:0;
  box-sizing: border-box;
}

.demo_container{
  padding:20px;
  width:800px;
  margin:0 auto;
  justify-content: center;
  align-items: center;
}

ul {
    list-style: none;
    counter-reset: nestedCounter;
  }
  ul li {
    counter-increment: nestedCounter;
    line-height: 1.4;
  }
  ul li:before {
    content: counters(nestedCounter, ".") " - ";
    font-weight: bold;
  }

</style>
</head>
```

```
<body>
<div class="demo_container">
  <h3> The counter() function enables to display
nested counters that have been generated by an
element and its parent/s. </h3>
  <ul>
    <li>Fruit
      <ul>
        <li>Apples
          <ul>
            <li>Green ones</li>
            <li>Red ones</li>
          </ul>
        </li>
        <li>Oranges
          <ul>
            <li>Small ones</li>
            <li>Big ones</li>
          </ul>
        </li>
      </ul>
    </li>
    <li>Vegetables
      <ul>
        <li>Carrots
          <ul>
            <li>Orange ones</li>
            <li>Purple ones</li>
          </ul>
        </li>
        <li>Potatoes
          <ul>
            <li>Fresh ones</li>
            <li>Rotten ones</li>
          </ul>
        </li>
      </ul>
    </li>
    <li>Fish
      <ul>
        <li>Big Ones
```

```
            <ul>
              <li>Mekong Catfish</li>
            </ul>
          </li>
          <li>Small Ones
            <ul>
              <li>Piranha</li>
              <li>Gold Fish</li>
              <li>Black Molly</li>
            </ul>
          </li>
        </ul>
      </li>
    </ul>
  </div>


  </body>
  </html>
```

The output of the code is given below:

The counter() function enables to display nested counters that have been generated by an element and its parent/s.

1 - Fruit
1.1 - Apples
1.1.1 - Green ones
1.1.2 - Red ones
1.2 - Oranges
1.2.1 - Small ones
1.2.2 - Big ones
2 - Vegetables
2.1 - Carrots
2.1.1 - Orange ones
2.1.2 - Purple ones
2.2 - Potatoes
2.2.1 - Fresh ones
2.2.2 - Rotten ones
3 - Fish
3.1 - Big Ones
3.1.1 - Mekong Catfish
3.2 - Small Ones
3.2.1 - Piranha
3.2.2 - Gold Fish
3.2.3 - Black Molly

CSS function (:counters()).

cubic-bezier()

The cubic-bezier() can be used with the transition timing function property to control how a transition will change speed over its duration. The property accepts an easing function which describes how the intermediate values used during a transition will be calculated.

The cubic-bezier() is an easing function that can provide a value for the transition-timing-function property. Some of the other various easing functions include ease-in, ease-out, linear, etc. However, the cubic-bezier() can be used to provide your own custom curve.

Example:

```
<!DOCTYPE html>
<html>
<head>
<style>

* {
  padding:0;
  margin:0;
  box-sizing: border-box;
}

.demo_container{
  padding:20px;
  width:800px;
  margin:0 auto;
  justify-content: center;
  align-items: center;
}
.ease {
    transition: width 1s ease;
  }
  .cubic-bezier {
    transition: width 1s
cubic-bezier(.63,.05,.43,1.7);
  }
  .ease:hover,
  .cubic-bezier: hover {
    width: 80%;
  }
```

```
    .ease, .cubic-bezier {
      background: orange;
      color: white;
      width: 90px;
      margin: 10px;
      padding: 10px;
    }
  </style>
  </head>
  <body>
  <div class="demo_container">
    <h3> The cubic-bezier() is used in CSS
  transitions to create a custom cubic Bézier curve.
  </h3>
    <div class="ease"> ease transition </div>
  <br>
    <div class="cubic-bezier">cubic-bezier()
  function </div>
    </div>
  </body>
  </html>
```

The output of the code is given below:

The cubic-bezier() is used in CSS transitions to create a custom cubic Bézier curve.

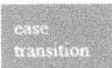

cubic-bezier() function

CSS function (cubic-bezier()).

drop-shadow()

It uses the drop-shadow() to apply a drop-shadow effect to an image. The CSS drop-shadow() is used with the filter property to add a drop-shadow effect to an image. A drop shadow is a blurred, offset version of the input image's alpha mask drawn in a particular color, composited below the image.

The CSS drop-shadow() accepts multiple arguments that determine the drop shadow's offset, its blur, and its color.

Example:

```
<!DOCTYPE html>
<html>
<head>
<style>

* {
  padding:0;
  margin:0;
  box-sizing: border-box;
}

.demo_container{
  padding:20px;
  width:800px;
  margin:0 auto;
  justify-content: center;
  align-items: center;
}

img{
  width:100%;
  height:250px;
}
.row{
  display: flex;
}
.col{

width:80%
}
.image-1 {
  filter: drop-shadow(5px 5px 10px gray);
  }
  .image-2 {
    filter: drop-shadow(0px 0px 10px orange);
    margin: 10px;
  }
```

```
    .image-3 {
      filter: drop-shadow(5px 5px 0 orange);
    }
    p{
      line-height: 2;
      font-size:20px
    }
</style>
</head>
<body>
<div class="demo_container">
  <h3> The contrast() functionused with the filter
property to adjust the contrast on given image.
</h3>
  <div class="row">
    <div class="col">
      <p> Normal Image </p>
      <img src="/images-1.jpg" alt="Sample
image">
      <p> drop-shadow(5px 5px 10px gray) Image
</p>
      <img class="image-1" src="/images-1.jpg"
alt="Sample image">
    </div>
    <div class="col">
      <p>drop-shadow(0px 0px 10px orange) Image
</p>
    <img class="image-2" src="/images-1.jpg"
alt="Sample image">
    <p>  drop-shadow(5px 5px 0 orange) Image
</p>
    <img class="image-3" src="/images-1.jpg"
alt="Sample image">
    </div>
  </div>
  </div>


</body>
</html>
```

The output of the code is given below:

The contrast() functionused with the filter property to adjust the contrast on given image.

Normal Image   drop-shadow(0px 0px 10px orange) Image

drop-shadow(5px 5px 10px gray) Image

drop-shadow(5px 5px 0 orange) Image

CSS function (:drop-shadow()).

ellipse()

The ellipse() function is a CSS basic shape value that's part of the CSS shapes module. The basic shapes such as ellipse() can be used as a value for properties such as shape-outside to control the flow of content (code) around the element, and clip-path to clip the element's contents to the basic shape. It means you can have text flowing around the element in the shape of an ellipse, have an image clipped to the shape of an ellipse, etc.

The syntax of the ellipse() function is as follows:

```
ellipse() = ellipse( [<shape-radius>{2}]? [at
<position>]? )
```

The syntax for <position> is:

```
<position> = [
  [ left | center | right | top | bottom | <length-
percentage> ]
|
  [ left | center | right | <length-percentage> ]
  [ top | center | bottom | <length-percentage> ]
|
  [ center | [ left | right ] <length-percentage>? ]
&&
  [ center | [ top | bottom ] <length-percentage>? ]
]
```

Example:

```
<!DOCTYPE html>
<html>
<head>
<style>

* {
  padding:0;
  margin:0;
  box-sizing: border-box;
}

.demo_container{
  width:800px;
  margin:0 auto;
  justify-content: center;
  align-items: center;
}

.section-1 {
  float: left;
  width: 200px;
  height: 150px;
  shape-outside: ellipse();
}
```

```
.section-2 {
  float: left;
  width: 200px;
  height: 150px;
  shape-outside: ellipse();
  background: yellow;
}
.section-3 {
  float: left;
  width: 200px;
  height: 150px;
  shape-outside: ellipse(30% 50%);
  background: yellow;
}
</style>
<div class="demo_container">
  <h1> The ellipse() is a CSS shape value that's
part of the CSS Shapes module circle(with no
background-color). </h1> <br>
  <section class="section-1"> </section>
  <p>Ornare quam viverra orci sagittis euvolutpat
odio. Viverra adipiscing at intellus integer
feugiat scelerisque.
    Adipiscing bibendum esttultricies integer quis
auctor. Massa tincidunt duiut ornare lectus sit
amet.
    Pellentesque elit egetg ravida cum sociis
natoque penatibus et. Sed vulputateodio ut enim
blandit volutpat maecenas volutpat.
    Purus viverra accumsan in nisl nisi.
Dignissim enimsit amet venenatis urna cursus
eget. Ornare arcu odiouut sem nulla pharetra diam
sit.
    Vitae justo eget magna fermentum iaculis.</p>
</div> <br> <br> <br>
<div class="demo_container">
  <h1> The ellipse() is a CSS shape value that's
part of the CSS Shapes module (with background-
color). </h1> <br>
  <section class="section-2"></section>
  <p>Ornare quam viverra orci sagittis euvolutpat
odio. Viverra adipiscing at intellus integer
feugiat scelerisque.
```

Adipiscing bibed ndum est ultricies integer
quis auctor. Massa ieltd tincidunt duiut ornare
lectus sit amet.

Pellentesque elit eget gravidacum sociis
natoque penatibus et. Sed vulputate odio ut enim
blandit volutpat maecenas volutpat.

Purus viverra accumsan inn nisl ntlisi.
Dignissim enim sit amet venenatis urnac ursus
eget. Ornare arcu odio utssem nulla pharetra diam
sit.

Vitae justo egetert magna fermentum
iaculis.</p>
</div>

```
<br> <br> <br>
<div class="demo_container">
  <h1> The ellipse() is a CSS shape value that's
part of the CSS Shapes module  shape-outside:
ellipse(30% 50%) </h1> <br>
  <section class="section-3"></section>
  <p>Ornare quam viverra orci sagittis euvolutpat
odio. Viverra adipiscing at in tellustels integer
feugiat scelerisque.
```

Adipiscing bibendum estultricies integer quis
auctor. Massa tinc idunt dui uttornare lectus sit
amet.

Pellentesque eliteget gravida cum sociis
natoque penatibus et. Sed vulput ateodio ut enim
blandit volutpat maecenas volutpat.

Purus viverra accumsan in nisl nisi.
Dignissieim sitamet venenatis urna cursus
eget. Ornare arcu odiout sem nulla pharetra
diam sit.

Vitae juso eget magna fermentum iaculis.</p>
</div>

```
</body>
</html>
```

The output of the code is given below:

The ellipse() is a CSS shape value that's part of the CSS Shapes module circle(with no backgroud-color).

Ornare quam viverra orci sagittis eu volutpat odio. Viverra adipiscing at in tellus integer feugiat scelerisque. Adipiscing bibendum est ultricies integer quis auctor. Massa tincidunt dui ut ornare lectus sit amet. Pellentesque elit eget gravida cum sociis natoque penatibus et. Sed vulputate odio ut enim blandit volutpat maecenas volutpat. Purus viverra accumsan in nisl nisi. Dignissim enim sit amet venenatis urna cursus eget. Ornare arcu odio ut sem nulla pharetra diam sit. Vitae justo eget magna fermentum iaculis.

The ellipse() is a CSS shape value that's part of the CSS Shapes module (with backgroud-color).

Ornare quam viverra orci sagittis eu volutpat odio. Viverra adipiscing at in tellus integer feugiat scelerisque. Adipiscing bibendum est ultricies integer quis auctor. Massa tincidunt dui ut ornare lectus sit amet. Pellentesque elit eget gravida cum sociis natoque penatibus et. Sed vulputate odio ut enim blandit volutpat maecenas volutpat. Purus viverra accumsan in nisl nisi. Dignissim enim sit amet venenatis urna cursus eget. Ornare arcu odio ut sem nulla pharetra diam sit. Vitae justo eget magna fermentum iaculis.

The ellipse() is a CSS shape value that's part of the CSS Shapes module shape-outside: ellipse(30% 50%)

Ornare quam viverra orci sagittis eu volutpat odio. Viverra adipiscing at in tellus integer feugiat scelerisque. Adipiscing bibendum est ultricies integer quis auctor. Massa tincidunt dui ut ornare lectus sit amet. Pellentesque elit eget gravida cum sociis natoque penatibus et. Sed vulputate odio ut enim blandit volutpat maecenas volutpat. Purus viverra accumsan in nisl nisi. Dignissim enim sit amet venenatis urna cursus eget. Ornare arcu odio ut sem nulla pharetra diam sit. Vitae justo eget magna fermentum iaculis.

CSS function (: ellipse ()).

Example:

```
<!DOCTYPE html>
<html>
<head>
<style>
```

```
*  {
   padding:0;
   margin:0;
   box-sizing: border-box;
}

.demo_container{
   width:800px;
   margin:0 auto;
   justify-content: center;
   align-items: center;
}

.section-1 {
  float: left;
  width: 200px;
  height: 150px;
 clip-path: ellipse();
 background-color: brown;
}

.section-2 {
  float: left;
  width: 200px;
  height: 150px;
  clip-path: ellipse(farthest-side closest-side at
50px 100px);
  background-color: brown;

}

</style>
<div class="demo_container">
  <h1> The ellipse() is a CSS shape value that's
part of the CSS Shapes module  clip-path:
ellipse()  </h1> <br>
  <section class="section-1"> </section>
  <p>Ornare quam viverra orci sagittis euvolutpat
odio. Viverra adipiscing attin tellus integer
feugiat scelerisque.
     Adipiscing bibendum est ultricies integer
quisauctor. Massa tincidunt dui ut ornar lectus
sit amet.
```

```
        Pellentesque eliteget gravida cumsociis
natoque penatibus et. Sed vul putate odio ut enim
blandit volutpat maecenas volutpat.
        Purus viverra accumsan in nisl nisi. Diggnissim
enim sit amet venenatis urna cursus eget. Ornare
arcu odioutsem nulla pharetra diam sit.
        Vitae justo eget magna fermentum iaculis.</p>
</br> <br> <br>


        <div class="demo_container">
        <h1> The ellipse() is a CSS shape value
that's part of the CSS Shapes module,  clip-path:
ellipse(farthest-side closest-side at 50px
100px);        </h1> <br>
        <section class="section-2"> </section>
        <p>Ornare quam viverra orci sagittis euuili
volutpat odio. Viverra adi piscing atin tellus
integer feugiat scelerisque.
        Adipiscing bibendum est ultricies
integquis auctor. Massa tincidunt duiiut ornare
lectus sit amet.
        Pelle ntesque elit eget gravida cummsociis
natoque penatibus et. Sed vulputate odiout enim
blandit volutpat maecenas volutpat.
        Purus viverra accumsan in nisl nisi.
Dignissim enim sit amet venenatis urcursus eget.
Ornare arcu odio ut sem nulla pharetra diamsit.
        Vitae justo egetm magna fermentum
iaculis.</p>
    </div>
</div>
</body>
</html>
```

filter()

The filter() function is used to apply a filter to an image. The filter() function allows you to apply filters to images. It's similar to the filter property, except that it's a function, and therefore can be used as a value itself. For example, you can use it with the background-image property to apply a filter to the background image.

The filter() function accepts two arguments. The first argument is the image and the second argument is a list of filter functions to apply to that image.

The filter functions that it accepts are the same as those accepted by the filter property. You can provide multiple filter functions if required.

Example:

```
<!DOCTYPE html>
<html>
<head>
<style>

* {
  padding:0;
  margin:0;
  box-sizing: border-box;
}

.demo_container{
  padding:20px;
  width:800px;
  margin:0 auto;
  justify-content: center;
  align-items: center;
}

div {
  background-size: cover;
  border: 1px solid black;
  height: 40vw;
  width: 40vw;
  margin: 0 2vw;
  float: left;
}
  .unfiltered {
    background-image: url('/images-1.jpg');
    height:200px;
    width:200px
  }
```

```
    .filtered {
        background-image: filter(url('/images-1.jpg'),
    hue-rotate(180deg));
        height:200px;
        width:200px
    }
    </style>
    </head>
    <body>
    <section class="demo_container">
        <h2> The filter() function is used to apply a
    filter to an image.</h2>
            <div class="unfiltered">
                Original Image
                </div>

                <div class="filtered">
                Filtered Image
                </div>
    </section>
    </body>
    </html>
```

The output of the code is given below:

The filter() function is used to apply a filter to an image.

 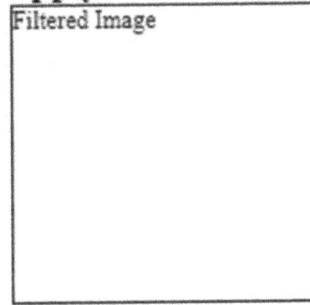

CSS function (filter()).

grayscale()

Use the grayscale() function to convert an image to grayscale. The CSS grayscale() is used with the filter property to convert an image to grayscale.

The grayscale() requires an argument to be passed to it. It determines the proportion of the conversion that's applied to the image. The argument can be either a % value or a number.

The syntax of the grayscale() function is as follows:

```
grayscale() = grayscale( [ <number> | <percentage> ] )
```

Example:

```
<!DOCTYPE html>
<html>
<head>
<style>

* {
  padding:0;
  margin:0;
  box-sizing: border-box;
}

.demo_container{
  padding:20px;
  width:800px;
  margin:0 auto;
  justify-content: center;
  align-items: center;
}

img{
  width:100%;
  height:250px;
}
.row{
  display: flex;
}
.col{

width:80%
}
.image-1 {
  filter: grayscale(100%);
  }
```

```css
  .image-2 {
    filter: grayscale(50%);


  }
  .image-3 {
    filter: grayscale(0.8);
  }
  p{
    line-height: 2;
    font-size:20px
  }
</style>
</head>
<body>
<div class="demo_container">
  <h2> The grayscale() function to convert an
image to grayscale. </h2>
  <div class="row">
    <div class="col">
      <p> Normal Image </p>
      <img src="/images-1.jpg" alt="Sample
image">
      <p> grayscale(100%) Image </p>
      <img class="image-1" src="/images-1.jpg"
alt="Sample image">
    </div>
    <div class="col">
      <p> grayscale(50%)  Image </p>
    <img class="image-2" src="/images-1.jpg"
alt="Sample image">
    <p>  grayscale(0.8)   Image </p>
    <img class="image-3" src="/images-1.jpg"
alt="Sample image">
    </div>
  </div>
  </div>


</body>
</html>
```

The output of the code is given below:

The grayscale() function to convert an image to grayscale.

Normal Image   grayscale(50%) Image

grayscale(100%) Image   grayscale(0.8) Image

CSS function (grayscale()).

hsl()

The hsl() function can be used to provide a color value when using CSS. It allows to specify that value by specifying the hue, saturation, and light components of the color.

HSL (stands for Hue Saturation Lightness) is a hue-based representation of the RGB color space of computer graphics.

The HSL model is considered to be more intuitive than the RGB model. The HSL model allows to select a base hue, and then adjust its saturation and lightness as desired. It accepts the HSL value as a parameter. The HSL value is provided as comma-separated three values that provide the hue, saturation, and light components, respectively. The hue can be identified by looking at the color circle.

Here's an example for displaying blue:

```
HSL(240, 100%, 50%)
```

How to Pick a Color

Here are few steps to picking a color. If you know the color, you can write it out in one step. However, if you are searching for a suitable color, following these steps can help others to find it in a methodical way.

1. Adjust the hue

2. Adjust the saturation

3. Adjust the lightness

Example:

```
<!DOCTYPE html>
<html>
<head>
<style>

* {
  padding:0;
  margin:0;
  box-sizing: border-box;
}

.demo_container{
  padding:20px;
  width:800px;
  margin:0 auto;
  justify-content: center;
  align-items: center;
}

img{
  width:100%;
  height:250px;
}

.image-1 {
background-color:  HSL(240, 100%, 50%)  ;
width:100%;
height:200px;
color:white;
padding:20px;
font-size:20px
  }
```

```
   .image-2 {
background-color: HSL(240, 10%, 50%)   ;
width:100%;
height:200px;
color:white;
padding:20px;
font-size:20px
   }
   .image-3 {
background-color:   HSL(240, 100%, 30%)   ;
width:100%;
height:200px;
color:white;
padding:20px;
font-size:20px
   }
  p{
    line-height: 2;
    font-size:20px
  }
</style>
</head>
<body>
<div class="demo_container">
  <h2> The hsl() function allows to specify a
color value by specifying the hue, saturation,
light components of the color. </h2>
  <div class="row">
     <p> hsl(240, 100%, 50%) Background </p>
     <div class="image-1"  alt="Sample image">
     Lorem ipsum dolorit amet, consectetur
adipiscing elit. Vivamus mauris dolor, semper
porta eros et, fermentum rhoncus erat.
     Cras facilisis mauris sit amet venenatis
aliquet. Suspendisse potent.
   </div>
     <p> hsl(240, 10%, 50%)     Background </p>
     <div class="image-2"  alt="Sample image">
     Lorem ipsum dolorit amet, consectetur
adipiscing elit. Vivamus mauris dolor, semper
porta eros et, fermentum rhoncus erat.
     Cras facilisis mauris sit amet venenatis
aliquet. Suspendisse potent.
   </div>
```

```
<p>  hsl(240, 100%, 30%)      Background </p>
<div class="image-3"  alt="Sample image">
     Lorem ipsum dolorit amet, consectetur
adipiscing elit. Vivamus mauris dolor, semper
porta eros et, fermentum rhoncus erat.
        Cras facilisis mauris sit amet venenatis
aliquet. Suspendisse potent.
        </div>
   </div>
  </div>
</body>
</html>
```

The output of the code is given below:

The hsl() function allows to specify a color value by specifying the hue, saturation, and light components of the color.

hsl(240, 100%, 50%) Background

Lorem ipsum dolor sit amet, consectetur adipiscing elit. Vivamus mauris dolor, semper porta eros et, fermentum rhoncus erat. Cras facilisis mauris sit amet venenatis aliquet. Suspendisse potent.

hsl(240, 10%, 50%) Background

Lorem ipsum dolor sit amet, consectetur adipiscing elit. Vivamus mauris dolor, semper porta eros et, fermentum rhoncus erat. Cras facilisis mauris sit amet venenatis aliquet. Suspendisse potent.

hsl(240, 100%, 30%) Background

Lorem ipsum dolor sit amet, consectetur adipiscing elit. Vivamus mauris dolor, semper porta eros et, fermentum rhoncus erat. Cras facilisis mauris sit amet venenatis aliquet. Suspendisse potent.

CSS function (HSL()).

hsla()

The CSS hsla() can be used to add transparency to a color when using the HSL model. It allows to specify a color value by specifying the hue, saturation, and light components of the color, as well as an alpha layer.

The hsla() is based on the HSL color model. HSL (that stands for Hue Saturation Lightness) is a hue-based representation of the RGB color.

The hsla() accepts the HSLA value as a parameter. It is provided as a comma-separated four values. The three HSL values (that provide the hue, saturation, and lightness components, respectively), and a fourth value, which provides the alpha channel.

Here's an example:

```
hsla(30, 100%, 50%, 0.5);
```

Example:

```
<!DOCTYPE html>
<html>
<head>
<style>

* {
  padding:0;
  margin:0;
  box-sizing: border-box;
}

.demo_container{
  padding:20px;
  width:800px;
  margin:0 auto;
  justify-content: center;
  align-items: center;
}

img{
  width:100%;
  height:250px;
}

.image-1 {
background-color: hsla(240, 100%, 50%, 1) ;
width:100%;
height:200px;
color:white;
padding:20px;
font-size:20px
  }
```

```css
  .image-2 {
background-color:hsla(240, 100%, 50%, 0.5)     ;
width:100%;
height:200px;
color:white;
padding:20px;
font-size:20px
   }
  .image-3 {
background-color:hsla(240, 100%, 50%, 0)     ;
width:100%;
height:200px;
color:white;
padding:20px;
font-size:20px;
border:1px solid black;
   }
  p{
    line-height: 2;
    font-size:20px
   }
</style>
</head>
<body>
<div class="demo_container">
  <h2>The hsla() function can be used to add
transparency to a color when using the HSL
model.  </h2>
  <div class="row">
      <p> hsla(240, 100%, 50%, 1) Background </p>
      <div class="image-1"  alt="Sample image">
        Lorem ipsum dolor sitamet, consectetur
adipiscing elit. Vivamus mauris dolor, semper
porta eros et, fermentum rhoncus erat.
        Cras facilisis mauris sit amet venenatis
aliquet. Suspendisse potent.
    </div>
      <p> hsla(240, 100%, 50%, 0.5)  Background </p>
      <div class="image-2"  alt="Sample image">
        Lorem ipsum dolor siamet, consectetur
adipiscing elit. Vivamus mauris dolor, semper
porta eros et, fermentum rhoncus erat.
```

```
            Cras facilisis mauris sit amet venenatis
aliquet. Suspendisse potent.
        </div>
        <p> hsla(240, 100%, 50%, 0)  Background </p>
        <div class="image-3"  alt="Sample image">
            Lorem ipsum dolor siamet, consectetur
adipiscing elit. Vivamus mauris dolor, semper
porta eros et, fermentum rhoncus erat.
            Cras facilisis mauris sit amet venenatis
aliquet. Suspendisse potent.
        </div>
    </div>
    </div>
</body>
</html>
```

The output of the code is given below:

The hsla() function can be used to add transparency to a color when using the HSL model.

hsla(240, 100%, 50%, 1) Background

Lorem ipsum dolor sit amet, consectetur adipiscing elit. Vivamus mauris dolor, semper porta eros et, fermentum rhoncus erat. Cras facilisis mauris sit amet venenatis aliquet. Suspendisse potent.

hsla(240, 100%, 50%, 0.5) Background

Lorem ipsum dolor sit amet, consectetur adipiscing elit. Vivamus mauris dolor, semper porta eros et, fermentum rhoncus erat. Cras facilisis mauris sit amet venenatis aliquet. Suspendisse potent.

hsla(240, 100%, 50%, 0) Background

CSS function (hsla()).

hue-rotate()

Use the hue-rotate() to apply a hue rotation on an image. The CSS hue-rotate() is used with the filter property to apply a hue rotation to an image. Where you specify an angle around the color circle that the input samples will be adjusted by. The hue-rotate() requires an argument to tell it how much to rotate the hue by.

The syntax of the hue-rotate() function is as follows:

```
hue-rotate() = hue-rotate( [ <angle> | <zero> ]? )
```

How Does Hue Rotation Work?

HSL (stands for Hue Saturation Lightness) is a hue-based representation of the RGB color space of computer graphics. The HSL model is considered to be more intuitive than the RGB model because, the HSL model allows to select a base hue, and then adjust its saturation and lightness as desired.

Example:

```
<!DOCTYPE html>
<html>
<head>
<style>

* {
  padding:0;
  margin:0;
  box-sizing: border-box;
}

.demo_container{
  padding:20px;
  width:800px;
  margin:0 auto;
  justify-content: center;
  align-items: center;
}

img{
  width:100%;
  height:250px;
}
```

```css
.row{
  display: flex;
}
.col{

width:80%
}
.image-1 {
  filter: hue-rotate(180deg);
  }
  .image-2 {
    filter: hue-rotate(-70deg);

  }

  p{
    line-height: 2;
    font-size:20px
  }
</style>
</head>
<body>
<div class="demo_container">
  <h3> The hue-rotate() function to apply a hue
rotation on an image. </h3>
  <div class="row">
    <div class="col">
      <p> Normal Image </p>
      <img src="/images-1.jpg" alt="Sample image">
      <p>  hue-rotate(180deg) Image </p>
      <img class="image-1" src="/images-1.jpg"
alt="Sample image">
    </div>
    <div class="col">
      <p> hue-rotate(-70deg)Image </p>
    <img class="image-2" src="/images-1.jpg"
alt="Sample image">

    </div>
  </div>
  </div>


</body>
</html>
```

The output of the code is given below:

The hue-rotate() function to apply a hue rotation on an image.

Normal Image   hue-rotate(-70deg)Image

hue-rotate(180deg) Image

CSS function (hue-rotate()).

inset()

The inset() is a CSS basic shape value that is part of the CSS shapes module. The shapes such as inset() can be used as a value for properties such as shape-outside to control the flow of content around the element and clip-path to clip the element's contents to the basic shape.

Example:

```
<!DOCTYPE html>
<html>
<head>
<style>

* {
  padding:0;
  margin:0;
  box-sizing: border-box;
}
```

```
.demo_container{
  width:800px;
  margin:0 auto;
  justify-content: center;
  align-items: center;
}

.section-1 {
  float: left;
  width: 200px;
  height: 150px;
  shape-outside: inset(50px);
}

.section-2 {
  float: left;
  width: 200px;
  height: 150px;
  shape-outside: inset(50px);
  background: yellow;

}

</style>
<div class="demo_container">
  <h1> The inset() function is a CSS basic shape
value part of the CSS Shapes module.  inset(50px)
no background   </h1> <br>
  <section class="section-1"> </section>
  <p>Ornare quam viverra orci sagittis euolutpat
odio. Viverra adipiscing atin tellus integer
feugiat scelerisque.
    Adipiscing bibendum est ultricies integer
quiauctor. Massa tincidunt duiut ornare lectus sit
amet.
    Pellente sque elit eget gravida cum
sociinatoque penatibus et. Sed vulputaoo ut enim
blandit volutpat maecenas volutpat.
    Purus viverra accumsan in nisl nisi.
Dignissim enisit amet venenatis urna cursus eget.
Ornare arcu odio uttsem nulla pharetra diam sit.
    Vitae justo eget magna fermentum iaculis.</p>
</br> <br> <br>
```

```
    <div class="demo_container">
      <h1> The inset() function is a CSS basic
shape value part of the CSS Shapes
module.  inset(50px) with background    </h1> <br>
      <section class="section-2"> </section>
      <p>Ornare quam viverra orci sagittis
euvolutpat odio. Viverra adipiscing atin tellus
integer feugiat scelerisque.
          Adipiscing bibendum est ultricies integer
quis auctor. Massa tincidunt duiut ornare lectus
sit amet.
          Pellentesque eleget gravida cum sociis
natoque penatibus et. Sed vulputate odio ut enim
blandit volutpat maecenas volutpat.
          Purus  accumsan in nislnisi. Dignissim
enim sit amet venenatis urna cur sus eget. Ornare
arcu odio utsem nulla pharetra diam sit.
          Vitae justo eget magna fermentum
iaculis.</p>
    </div>
</div>
</body>
</html>
```

The output of the code is given below:

The inset() function is a CSS basic shape value part of the CSS Shapes module. inset(50px) no background

Ornare quam viverra orci sagittis eu volutpat odio. Viverra adipiscing at in tellus integer feugiat scelerisque. Adipiscing bibendum est ultricies integer quis auctor. Massa tincidunt dui ut ornare lectus sit amet. Pellentesque elit eget gravida cum sociis natoque penatibus et. Sed vulputate odio ut enim blandit volutpat maecenas volutpat. Purus viverra accumsan in nisl nisi. Dignissim enim sit amet venenatis urna cursus eget. Ornare arcu odio ut sem nulla pharetra diam sit. Vitae justo eget magna fermentum iaculis.

The inset() function is a CSS basic shape value part of the CSS Shapes module. inset(50px) with background

Ornare quam viverra orci sagittis eu volutpat odio. Viverra adipiscing at in tellus integer feugiat scelerisque. Adipiscing bibendum est ultricies integer quis auctor. Massa tincidunt dui ut ornare lectus sit amet. Pellentesque elit eget gravida cum sociis natoque penatibus et. Sed vulputate odio ut enim blandit volutpat maecenas volutpat. Purus viverra accumsan in nisl nisi. Dignissim enim sit amet venenatis urna cursus eget. Ornare arcu odio ut sem nulla pharetra diam sit. Vitae justo eget magna fermentum iaculis.

CSS function (inset()).

linear-gradient()

The linear-gradient() function allows to create a linear-gradient using CSS. CSS gradients allow to apply multiple background colors to an element that blend from one color to the next.

To create a linear-gradient, use the linear-gradient() as a value to any property that accepts images (such as background-image, background, or border-image properties). Linear-gradients use these properties to build an image of the specified gradient.

These are gradients that run in a straight line. Linear-gradients can go in any direction such as up, down, sideways, 50 degrees, 73 degrees, or any other angle you like.

These gradients are created by establishing an axis called the gradient line. The color stops are placed along that gradient to create a gradient that runs between each color.

How to Set the Angle of a Gradient

You can also set the angle so that it runs along that angle. You can specify with an angle value (e.g., 45deg, 90deg, and 180deg) or with the keywords for specifying the angle (such as to top, to bottom, to top left, etc).

Example:

```
<!DOCTYPE html>
<html>
<head>
<style>

* {
  padding:0;
  margin:0;
  box-sizing: border-box;
}

.demo_container{
  padding:20px;
  width:900px;
```

```
  margin:0 auto;
  justify-content: center;
  align-items: center;
}


.row{
  display: flex;
  flex-direction:row;
}
.col{
  height:100px;
width:100%
}
.image-1 {
  background: linear-gradient(purple, yellow);
    color: white;
    padding: 30px;
    height: 300px;
  }
  .image-2 {
    background: linear-gradient(to bottom right,
purple, yellow);
    color: white;
    padding: 30px;
    height: 300px;
  }
  .image-3{
    background: linear-gradient(to right, red 5%,
green 0%, yellow 5%);
    color: white;
    padding: 30px;
    height: 300px;
  }
  .image-4{
    background: linear-gradient(to right, red,
green, yellow);
    color: white;
    padding: 30px;
    height: 300px;
  }
```

```
   p{
     line-height: 2;
     font-size:20px;
   margin-bottom:30px
   }
</style>
</head>
<body>
<div class="demo_container">
  <h2>The linear-gradient() function allows you to
create a linear gradient using CSS.</h2>
   <div class="row">
     <div class="col">
       <p> Default Angle,  linear-gradient(purple,
yellow) </p>
       <div class="image-1"> </div>
       <p> Two color, linear-gradient(to bottom
right, purple, yellow) </p>
       <div class="image-2"> </div>
     </div>
     <div class="col">
       <p> Positioning the Color, linear-
gradient(to right, red 5%, green 30%, yellow 75%)
</p>
       <div class="image-3"> </div>
     <p> Three color, linear-gradient(to right,
red, green, yellow) </p>
     <div class="image-4"> </div>

     </div>
   </div>
   </div>


</body>
</html>
```

The output of the ode is given below:

The contrast() functionused with the filter property to adjust the contrast on given image.

Default Angle, linear-gradient(purple, yellow)   Positioning the Color, linear-gradient(to right, red

5%, green 30%, yellow 75%)

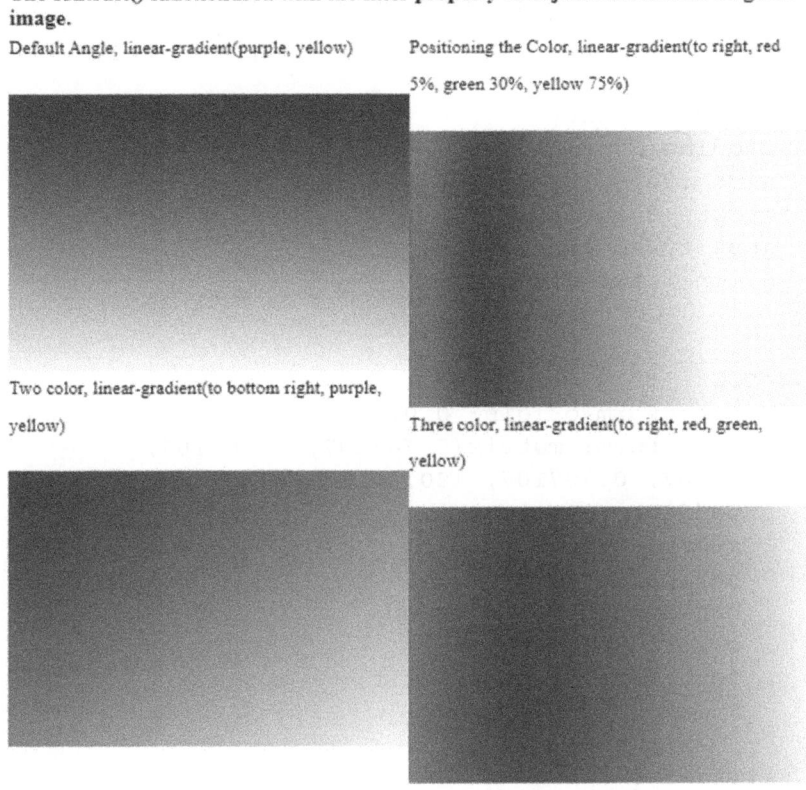

Two color, linear-gradient(to bottom right, purple,

yellow)   Three color, linear-gradient(to right, red, green,

yellow)

CSS function linear-gradient().

matrix()

The CSS matrix() can be used with CSS transforms to style elements in a two-dimensional space.

The matrix() is an alternative to the two-dimensional transform functions rotate(), skew(), scale(), and translate().

Example:

```
<!DOCTYPE html>
<html>
<head>
<style>
```

```css
* {
  padding:0;
  margin:0;
  box-sizing: border-box;
}

.demo_container{
  padding:20px;
  width:900px;
  margin:0 auto;
  justify-content: center;
  align-items: center;
}

.matrix-1{
    transform-origin: 0 0;
    transform: matrix(0.707107, 0.707107,
-0.707107, 0.707107, 150, 0);
    padding: 20px;
    width: 300px;
    height:300px;
    background: lime green;
    color: white;
    font-family: sans-serif;
}

.matrix-2 {
    transform-origin: 0 0;
    transform: matrix(2, 0, 0, 2, 0, 0);
    padding: 20px;
    width: 120px;
    background: lime green;
    color: white;
    font-family: sans-serif;
}
.row{
  display: flex;

}
.col{
  width:100%
}
</style>
</head>
```

```
<body>
<div class="demo_container">
  <h1>The matrix() function can be used with CSS
transforms to style elements in a two-dimensional
space. </h1>
  <div class="row">
    <div class="col">
      <div class="matrix-1">  Scaled and moved
with <code>matrix() function =      transform:
matrix(0.707107, 0.707107, -0.707107, 0.707107,
150, 0);  </code> </div>
    </div>
    <div class="col">
    <div class="matrix-2">  Scaled and moved with
<code>matrix() function =    transform: matrix(2,
0, 0, 2, 0, 0);
  </div>

  </div>
  </code>
  </div>
  </div>
</body>
</html>
```

The output of the code is given below:

The matrix() function can be used with CSS transforms to style elements in a two-dimensional space.

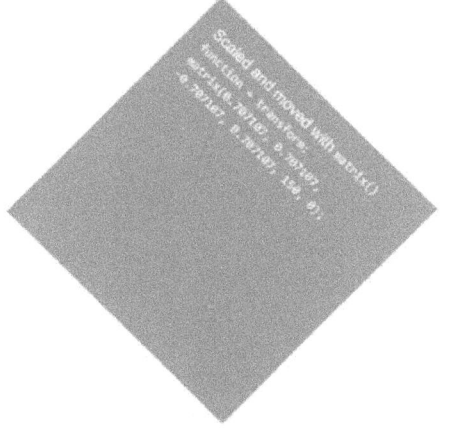

CSS function- matrix().

matrix3d()

The CSS matrix3d() can be used with CSS transforms to style elements in a three-dimensional space. The matrix3d() is an alternative to the three-dimensional transform functions rotate3d(), rotateX(), rotateY(), rotateZ(), translate3d(), translateZ(), scale3d(), scaleZ(), and perspective().

Example:

```
<!DOCTYPE html>
<html>
<head>
<style>

* {
  padding:0;
  margin:0;
  box-sizing: border-box;
}

.demo_container{
  padding:20px;
  width:900px;
  margin:0 auto;
  justify-content: center;
  align-items: center;
}

.matrix-1{
    transform-origin: 0 0;
    padding: 20px;
    width: 300px;
    height:300px;
    background: lime green;
    color: white;
    font-family: sans-serif;
}

.matrix-2 {
    transform-origin: 0 0;
    transform: matrix3d(0.583333, 0.186887, 0.9044,
0, -0.52022, 0.833333, 0.186887, 0, -0.623773,
-0.52022, 0.583333, 0, 0, 0, 0, 1);    padding:
20px;
```

```
      width: 300px;
      height:250px;
      background: lime green;
      color: white;
      font-family: sans-serif;

   }
.row{
   display: flex;

}
.col{
   width:100%
}
</style>
</head>
<body>
<div class="demo_container">
   <h1>The CSS matrix3d() function can be used with
CSS transforms to style elements in a three-
dimensional space. </h1>
   <div class="row">
     <div class="col">
       <div class="matrix-1">    Scaled and moved
with <code>matrix() function =      transform:
matrix(0.707107, 0.707107, -0.707107, 0.707107,
150, 0);  </code> </div>
     </div>
     <div class="col">
       <div class="matrix-2">   Scaled and moved with
<code>matrix() function =       transform: matrix(2,
0, 0, 2, 0, 0);
   </div>

   </div>
   </code>
   </div>

   </div>


</body>
</html>
```

The output of the code is given below:

The CSS matrix3d() function can be used with CSS transforms to style elements in a three-dimensional space.

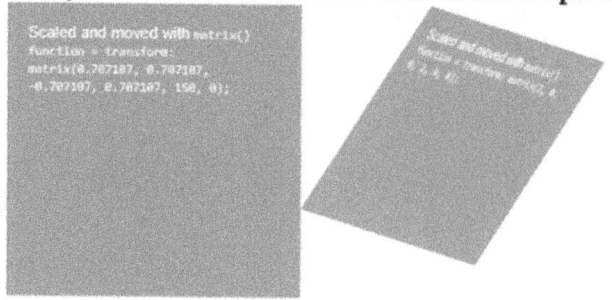

CSS function-matrix3d().

opacity()

Use the opacity() function to make an image partially or completely transparent. The opacity() is used with the filter property to apply transparency to the samples in an image. The opacity() requires an argument to be passed to it. Also, it determines the proportion of the transparency that's applied to the image. The argument can be either a value or a number.

Example:

```
<!DOCTYPE html>
<html>
<head>
<style>

body {
    padding:0;
    margin:0;
    box-sizing: border-box;
    background: gold;

}

.demo_container{
    padding:20px;
    width:900px;
    margin:0 auto;
```

```css
    justify-content: center;
    align-items: center;
}
.row{
  display: flex;
  flex-wrap:wrap ;

}
.box-1{
    transform-origin: 0 0;
    padding: 20px;
    width: 300px;
    height:300px;
    background-color: green;
    filter: opacity(10%);
    color: white;
    font-family: sans-serif;
}

.box-2{
    transform-origin: 0 0;
    padding: 20px;
    width: 300px;
    height:300px;
    background-color: green;
    filter: opacity(40%);
    color: white;
    font-family: sans-serif;
}

.box-3{
    transform-origin: 0 0;
    padding: 20px;
    width: 300px;
    height:300px;
    background-color: green;
    filter: opacity(70%);
    color: white;
    font-family: sans-serif;
}
.box-4{
    transform-origin: 0 0;
    padding: 20px;
    width: 300px;
```

```
        height:300px;
        background-color: green;
        filter: opacity(90%);
        color: white;
        font-family: sans-serif;
}
</style>
</head>
<body>
<div class="demo_container">
   <h1>The CSS matrix3d() function can be used with
CSS transforms to style elements in a three-
dimensional space. </h1>
   <div class="row">
      <div class="box-1">  opacity(10%)   </div>
      <div class="box-2">  opacity(40%) </div>
      <div class="box-3">  opacity(70%) </div>
      <div class="box-4">  opacity(90%) </div>
   </div>
   </div>
</body>
</html>
```

The output of the code is given below:

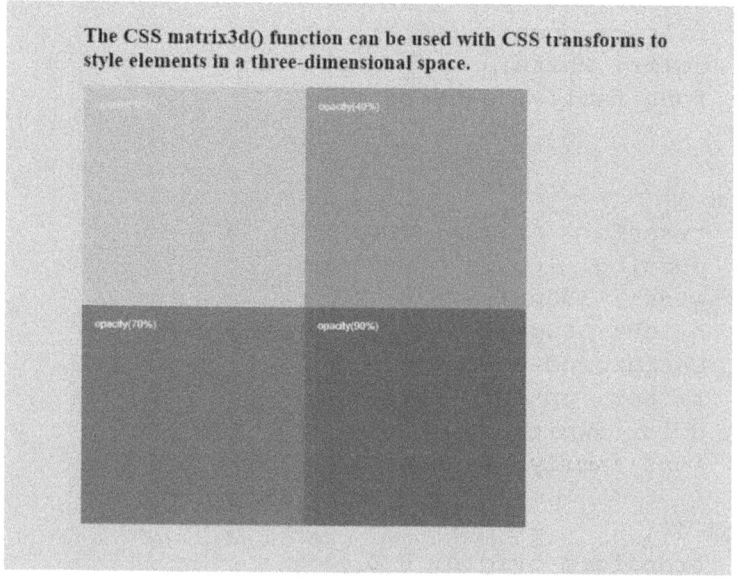

CSS function-opacity().

perspective()

The perspective() function defines the distance between the z=0 plane and the user in order to give the 3D-positioned element some perspective.

The perspective() function works like this:

```
perspective(l)
```

The l parameter specifies the distance from the ones to the z=0 plane. The argument is also specified as a length value (e.g., 1px, 1vw, etc). The greater the length, the less pronounced the "3D effect" will be.

Example:

```
<!DOCTYPE html>
<html>
<head>
<style>

* {
  padding:0;
  margin:0;
  box-sizing: border-box;
}

.demo_container{
  padding:20px;
  width:800px;
  margin:0 auto;
  justify-content: center;
  align-items: center;
}

img{
  width:100%;
  height:250px;
}
.row{
  display: flex;
}
.col{

width:80%
}
```

```css
.image-1 {
  transform: perspective(400px) rotateY(70deg);
  width:100px;
  height:100px
  }
  .image-2 {
    transform: perspective(200px) rotateY(60deg);
    width:100px;
  height:100px

  }
  .image-3 {
    transform: perspective(130px) rotateY(-60deg);
    width:100px;
  height:100px
  }
  p{
    line-height: 2;
    font-size:20px
  }
</style>
</head>
<body>
<div class="demo_container">
  <h2> It defines the distance between the z=0
plane and the user in order to give the
3D-positioned element some perspective </h2>
  <div class="row">
    <div class="col">
      <p> Normal Image </p>
      <img src="/images-1.jpg" alt="Sample
image">
      <p> perspective(400px) rotateY(70deg)   Image
</p>
      <img class="image-1" src="/images-1.jpg"
alt="Sample image">
    </div>
    <div class="col">
      <p> perspective(200px) rotateY(60deg) Image
</p>
    <img class="image-2" src="/images-1.jpg"
alt="Sample image">
```

```
    <p>  perspective(130px) rotateY(-60deg)
Image </p>
    <img class="image-3" src="/images-1.jpg"
alt="Sample image">
    </div>
  </div>
  </div>


</body>
</html>
```

The output of the code is given below:

It defines the distance between the z=0 plane and the user in order to give the 3D-positioned element some perspective

Normal Image

perspective(200px) rotateY(60deg) Image

perspective(130px) rotateY(-60deg) Image

perspective(400px) rotateY(70deg) Image

CSS function (perspective()).

radial-gradient()

The radial-gradient() function allows to create a radial gradient using CSS.

CSS gradients allow to apply multiple background colors to an element that blend from one color to the next.

Example:

```
<!DOCTYPE html>
<html>
<head>
<style>

* {
  padding:0;
  margin:0;
  box-sizing: border-box;
}

.demo_container{
  padding:20px;
  width:900px;
  margin:0 auto;
  justify-content: center;
  align-items: center;
}

.row{
  display: flex;
  flex-direction:row;
}
.col{
  height:100px;
width:100%
}
.image-1 {
  background: radial-gradient(yellow, red);
    color: white;
    padding: 30px;
    height: 300px;
    }
  .image-2 {
    background: radial-gradient(ellipse, black,
lime);
    color: white;
    padding: 30px;
    height: 300px;
  }
```

```
    .image-3{
      background: radial-gradient(circle, black,
lime);
      color: white;
      padding: 30px;
      height: 300px;
    }
    .image-4{
      background: radial-gradient(at bottom right,
yellow, red);
      color: white;
      padding: 30px;
      height: 300px;
    }
    p{
      line-height: 2;
      font-size:20px;
    margin-bottom:30px
    }
</style>
</head>
<body>
<div class="demo_container">
  <h2>The linear-gradient() function allows
you to create a linear gradient using
CSS.</h2>
  <div class="row">
    <div class="col">
      <p> Default Angle, radial-gradient(yellow,
red) </p>
      <div class="image-1"> </div>
      <p> Ellipse shape, radial-gradient(ellipse,
black, lime) </p>
      <div class="image-2"> </div>
    </div>
    <div class="col">
      <p> Circle, radial-gradient(circle, black,
lime)</p>
      <div class="image-3"> </div>
```

```
    <p> Gradient Position, radial-gradient(at
bottom right, yellow, red) </p>
    <div class="image-4"> </div>
    </div>
  </div>
  </div>
</body>
</html>
```

The output of the code is given below:

The radial-gradient() function allows you to create a radial gradient using CSS.

Default Angle, radial-gradient(yellow, red)   Circle ,radial-gradient(circle, black, lime)

Ellipse shape, radial-gradient(ellipse, black, lime)   Gradient Position, radial-gradient(at bottom right, yellow, red)

CSS Function (radial-gradient() ())

repeating-linear-gradient()

The CSS repeating-linear-gradient() allows to create a linear gradient that repeats over and over again infinitely in both directions. The gradients allow to apply multiple background colors to an element that blend from one color to the next. These gradients are linear gradients where the color stops are repeated infinitely in both directions. It is repeating linear gradients work the same way that (are non-repeating) linear gradients work.

Syntax: The repeating-linear-gradient() function accepts the following values:

- angle: It specifies an angle for the direction of the gradient. For example, 45deg.

The following keywords specify to point the gradient-line given below:

- to top
- to right
- to bottom
- to left

The following keywords specify to point the gradient-line given below:

- to top left
- to top right
- to bottom right
- to bottom left
- color-stop

Example:

```
<!DOCTYPE html>
<html>
<head>
<style>

* {
  padding:0;
  margin:0;
  box-sizing: border-box;
}
```

```
.demo_container{
  padding:20px;
  width:900px;
  margin:0 auto;
  justify-content: center;
  align-items: center;
}

.row{
  display: flex;
  flex-direction:row;
}
.col{
  height:100px;
width:100%
}

.image-1 {
  background: repeating-linear-gradient(gold 15%,
orange 30%);
    color: white;
    padding: 30px;
    height: 300px;
   }
  .image-2 {
    background: repeating-linear-gradient(to top
right, gold 16%, orange 30%);
    color: white;
    padding: 30px;
    height: 300px;
  }
  .image-3{
    background: repeating-linear-gradient(to  top
right, orange, gold 16%, orange 30%);
    color: white;
    padding: 30px;
    height: 300px;
  }
  .image-4{
    background: repeating-linear-gradient(155deg,
gold, gold 50px, orange 60px, orange 120px);
    color: white;
```

```
    padding: 30px;
    height: 300px;
  }
  p{
    line-height: 2;
    font-size:20px;
    margin-bottom:30px
  }


</style>
</head>
<body>
<div class="demo_container">
  <h2>The repeating-linear-gradient() function
allows you to create a linear gradient that
repeats over and over again infinitely in both
directions.</h2>
  <div class="row">
    <div class="col">
      <p> Default Angle, radial-gradient(yellow,
red) </p>
      <div class="image-1"> </div>
      <p> Angle of a Repeating, repeating-linear-
gradient( top right, orange, gold 15%, orange 30%)
</p>
      <div class="image-2"> </div>
    </div>
    <div class="col">
      <p> Smooth Transitions, repeating-linear-
gradient( top right, orange, gold 15%, orange 30%)
</p>
      <div class="image-3"> </div>
    <p>Creating Stripes, repeating-linear-
gradient(165deg, gold, gold 60px, orange 60px,
orange 120px)</p>
      <div class="image-4"> </div>
    </div>
  </div>
  </div>
</body>
</html>
```

The output of the code is given below:

The repeating-linear-gradient() function allows you to create a linear gradient that repeats over and over again infinitely in both directions.

Default Angle, radial-gradient(yellow, red)

Smooth Transitions ,repeating-linear-gradient(top right, orange, gold 15%, orange 30%)

Angle of a Repeating, repeating-linear-gradient(top right, orange, gold 15%, orange 30%)

Creating Stripes, repeating-linear-gradient(165deg, gold, gold 60px, orange 60px, orange 120px)

CSS function (repeating-linear -gradient() ()).

repeating-radial-gradient()

The CSS repeating-radial-gradient() allows to create a radial gradient that repeats over and over again. CSS gradients allow you to apply multiple bg colors to an element that blend from single color to the next. Repeating radial gradients are gradients where the color stops are repeated infinitely. Radial gradients have a circular or elliptical shape. When creating a radial gradient, indicate the center of the gradient as well as the size and shape.

Syntax: This function uses the same syntax as the radial-gradient() function, which is as follows:

```
<radial-gradient> = radial-gradient(
  [ [ <shape> || <size> ] [ at <position> ]?,  |
    at <position>,
  ]?
  <color-stop> [,  <color-stop> ]+
)
```

The repeating-radial-gradient() function accepts the following values:

- shape: It specifies the shape of the gradient. It can be either circle or ellipse.

- size: It specifies the size of the ending shape. The values are given below:

  - closest side

  - farthest side

  - closest corner

  - farthest corner

  - length (for circles)

  - [length | percentage] (for ellipses)

- position: it specifies where the gradient should be located. The values are given below:

  - top

  - right

  - bottom

  - left

  - center

Example:

```
<!DOCTYPE html>
<html>
<head>
<style>

* {
  padding:0;
  margin:0;
  box-sizing: border-box;
}

.demo_container{
  padding:20px;
  width:900px;
  margin:0 auto;
  justify-content: center;
  align-items: center;
}

.row{
  display: flex;
  flex-wrap: wrap;
  flex-direction: row;
}
.col{
width:50%
}
p{
  height:50px
}
.image-1 {
  background: repeating-radial-gradient(ellipse,
green 20%, lime 40%) ;
    color: white;
    padding: 30px;
    height: 300px;
    }
  .image-2 {
    background: repeating-radial-gradient(circle,
green 20%, lime 40%);
    color: white;
```

```
      padding: 30px;
      height: 300px;
  }
   .image-3{
      background:repeating-radial-gradient(circle at
top left, green 30%, lime 40%);
      color: white;
      padding: 30px;
      height: 300px;
  }

   .image-4{
      background: repeating-radial-gradient(circle
at top left, lightgreen 10%, green 45%, lime
60%);
      color: white;
      padding: 30px;
      height: 300px;
  }
   .image-5{
      background: repeating-radial-gradient(circle
at top left, lime, green 30%, lime 40%);
      color: white;
      padding: 30px;
      height: 300px;
  }
   .image-6{
      background: repeating-radial-gradient(closest-
side at 25px 35px, orange 25%, gold 40%);
      background-size:50px 50px;
      height: 200px;
      color: white;
      padding: 30px;
      height: 300px;
  }
  p{
      line-height: 2;
      font-size:20px;
   margin-bottom:30px
  }

</style>
</head>
```

```
<body>
<div class="demo_container">
  <h2>CSS repeating radial gradient() function
allows you to create a radial gradient that
repeats over and over again infinitely.</h2>
  <div class="row">
    <div class="col">
      <p> Ellipse, repeating-radial-
gradient(ellipse, green 20%, lime 40%) </p>
      <div class="image-1"> </div>
      <p>Circle, repeating radial gradient(circle,
green 20%, lime 40%) </p>
      <div class="image-2"> </div>
      <p>Gradient Position, repeating-radial-
gradient(circle at top left, green 20%, lime
40%)</p>
      <div class="image-3"> </div>

    </div>

    <div class="col">
      <p>Three Color Stops, repeating-radial-
gradient(circle at top left, lightgreen 20%, green
55%, lime 60%)</p>
    <div class="image-4"> </div>
      <p> Smooth Transitions, repeating-radial-
gradient(circle at top left, lime, green 20%, lime
40%)</p>
      <div class="image-5"> </div>
      <p>Patterns with Gradients, repeating radial
gradient(closest-side at 25px 35px, orange 15%,
gold 40%)</p>
      <div class="image-6"> </div>

    </div>


  </div>
  </div>


</body>
</html>
```

The output of the code is given below:

CSS repeating-radial-gradient() function allows you to create a radial gradient that repeats over and over again infinitely.

Ellipse, repeating-radial-gradient(ellipse, green 20%, Three Color Stops, repeating-radial-gradient(circle at

lime 40%) top left, lightgreen 20%, green 55%, lime 60%)

Circle, repeating-radial-gradient(circle, green 20%, Smooth Transitions, repeating-radial-gradient(circle

lime 40%) at top left, lime, green 20%, lime 40%)

Gradient Position, repeating-radial-gradient(circle at Patterns with Gradients, repeating-radial-

top left, green 20%, lime 40%) gradient(closest-side at 25px 35px, orange 15%, gold

40%)

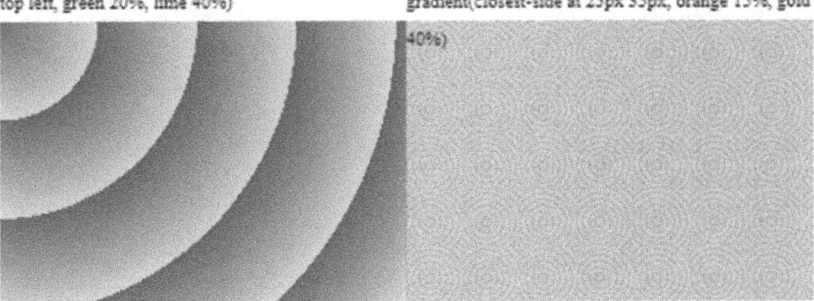

CSS function (repeating-radial -gradient()).

rgb()

The rgb() function can be used to provide a value when using CSS. It allows to specify an RGB color value by specifying the red, green, and blue channels directly.

RGB (stands for Red, Green, Blue) is a color model in which red, green, and blue light can be added together to reproduce a color.

The RGB() accepts the RGB value as a parameter. The RGB function is provided as comma-separated three values. It provides red, green, and blue hues, respectively.

Each of the three values can be provided as an integer or as a percentage.

Example:

```
RGB(255,0,0)
RGB(100%,0%,0%)
```

We can use the property on any element.

Example:

```
<!DOCTYPE html>
<html>
<head>
<style>

* {
  padding:0;
  margin:0;
  box-sizing: border-box;
}

.demo_container{
  padding:20px;
  width:900px;
  margin:0 auto;
  justify-content: center;
  align-items: center;
}

.row{
  display: flex;
  flex-wrap: wrap;
  flex-direction: row;
}
```

```
.col{
width:50%
}
p{
  height:50px
}
.image-1 {
  background: RGB(255,69,0);
    color: white;
    padding: 30px;
    height: 100px;
   }
  .image-2 {
    background: RGB(0,255,0);
    color: white;
    padding: 30px;
    height: 100px;
  }
  .image-3{
    background: RGB(0,0,128);
    color: white;
    padding: 30px;
    height: 100px;
  }

  .image-4{
    background: RGB(128,128,0);
    color: white;
    padding: 30px;
    height: 100px;
  }
  .image-5{
    background: RGB(255,0,0);
    color: white;
    padding: 30px;
    height: 100px;
  }
  .image-6{
    background: RGB(128,0,128);
    height: 100px;
    color: white;
    padding: 30px;
  }
```

```css
.image-7{
  background:  RGB(255,0,70);
  height: 100px;
  color: white;
  padding: 30px;
}

.image-8{
  background: RGB(230,255,0);
  height: 100px;
  color: white;
  padding: 30px;
}
p{
  line-height: 2;
  font-size:24px;
margin-bottom:10px
}

</style>
</head>
<body>
<div class="demo_container">
  <h2>The rgb() function can be used to provide a
color value when using CSS</h2>
  <div class="row">
    <div class="col">
      <p>   rgb(255,69,0) </p>
      <div class="image-1"> </div>
      <p> rgb(0,255,0) </p>
      <div class="image-2"> </div>
      <p> rgb(0,0,128) </p>
      <div class="image-3"> </div>
      <p>   rgb(230,255,0) </p>
      <div class="image-8"> </div>

    </div>

    <div class="col">
      <p> rgb(128,128,0) </p>
    <div class="image-4"> </div>
      <p> rgb(255,0,0) </p>
      <div class="image-5"> </div>
```

```
<p> rgb(128,0,128)</p>
<div class="image-6"> </div>
<p>  rgb(255,0,70) </p>
<div class="image-7"> </div>

    </div>
  </div>
  </div>
</body>
</html>
```

The output of the code is given below:

The rgb() function can be used to provide a color value when using CSS

rgb(255,69,0) rgb(128,128,0)

rgb(0,255,0) rgb(255,0,0)

rgb(0,0,128) rgb(128,0,128)

rgb(230,255,0) rgb(255,0,70)

CSS function (RGB ()).

rotate()

The CSS rotate() is used to rotate elements in two-dimensional space. The rotate() rotates an element based on the angle that you provide as an argument. You can also provide the angle using any valid CSS angle value.

There are various properties used in rotate such as:

1. Degrees

2. Gradians

3. Turns

Example:

```
<!DOCTYPE html>
<html>
<head>
<style>

* {
  padding:0;
  margin:0;
  box-sizing: border-box;
}

.demo_container{
  padding:20px;
  width:900px;
  margin:0 auto;
  justify-content: center;
  align-items: center;
}

.row{
  display: flex;
  flex-wrap: wrap;
  flex-direction: row;
}
.col{
width:50%
}
p{
  height:50px
}
.image-1 {
  background: RGB(255,69,0);
    color: white;
```

```
  padding: 30px;
  height: 50px;
  width:100px


 }
.image-2 {
  background: RGB(0,255,0);
  transform: rotate(15deg);
  color: white;
  padding: 30px;
  height: 50px;
  width:100px
  }
.image-3{
  transform: rotate(20grad);
  background: rgb(230,255,0);
  color: white;
  padding: 30px;
  height: 50px;
  width:100px
  }

.image-4{
  background: RGB(128,128,0);
  transform: rotate(.4turn);
  color: white;
  padding: 30px;
  height: 50px;
  width:100px
  }
.image-5{
  background: #ff0000;
  transform: rotate(-15deg);
  color: white;
  padding: 30px;
  height: 50px;
  width: 100px
  }
.image-6{
  background: RGB(128,0,128);
  transform-origin: 90% 90%;
  transform: rotate(15deg);
```

```
    height: 50px;
    width:100px;
    color: white;
    padding: 30px;
  }
  p{
    line-height: 2;
    font-size:24px;
   margin-bottom:10px
   }


</style>
</head>
<body>
<div class="demo_container">
  <h2>The rotate() function is used to
rotate elements in a two-dimensional
space.</h2>
  <div class="row">
    <div class="col">
      <p>  Not rotated </p>
      <div class="image-1"> </div>
      <p> Degrees (rotate(15deg)) </p>
      <div class="image-2"> </div>
      <p> Gradians rotate(20grad) </p>
      <div class="image-3"> </div>
    </div>
    <div class="col">
      <p>Turns (rotate(.4turn)) </p>
    <div class="image-4"> </div>
      <p> Negative Values (rotate(-15deg))</p>
      <div class="image-5"> </div>
      <p> Adding transform-origin </p>
      <div class="image-6"> </div>
    </div>
  </div>
  </div>
</body>
</html>
```

The output of the code is given below:

The rotate() function is used to rotate elements in a two-dimensional space.

Not rotated

Turns (rotate(.4turn))

Degrees (rotate(15deg))

Negative Values (rotate(-15deg))

Gradians rotate(20grad)

Adding transform-origin

CSS function (rotate()).

rotate3d()

The rotate3d() function is used to rotate elements in three-dimensional space. The rotate3d() function rotates the element along the x, y, and z axes using the angle provided as an argument.

The syntax of the rotate3d() function is as follows:

```
rotate3d() = rotate3d( <number>, <number>, <number>,
<angle> ).
```

The first three parameters describe the [x, y, z] direction vectors. The fourth parameter specifies the angle to be used. It works like this:
rotate3d(x, y, z, a)

- x: It describes the x-coordinate of the vector denoting the axis of rotation.

- y: It describes the y-coordinate of the vector denoting the axis of rotation.

- z: It describes the z-coordinate of the vector denoting the axis of rotation.

- a: It represents the angle of the rotation. A positive results in a clockwise rotation, and a negative results in a counter-clockwise rotation.

- deg: It represents as degrees. There are 360 degrees of angle in a full circle.

- grad: It is gradians, also known as "gons" or "grades." There are 400 gradians in a whole circle.

- rad: It is radians. There are 2ï€ radians in a circle.

- turn: It is turns. There is one turn in a circle.

Example:

```
<html>
<head>
<style>

* {
  padding:0;
  margin:0;
  box-sizing: border-box;
}

.demo_container{
  padding:20px;
  width:900px;
  margin:0 auto;
  justify-content: center;
  align-items: center;
}

.row{
  display: flex;
  flex-wrap: wrap;
  flex-direction: row;
}
.col{
width:50%
}
p{
  height:50px
}
.image-1 {
  background: RGB(255,69,0);
    color: white;
    padding: 30px;
```

```
    height: 50px;
    width:100px
   }
 .image-2 {
   background: RGB(0,255,0);
  transform: rotate3d(1, 0, 0, 60deg);
   color: white;
   padding: 30px;
   height: 50px;
   width:100px
 }

 .image-4{
   background: RGB(128,128,0);
   transform: rotate3d(1, 5, 1, 60deg);
   color: white;
   padding: 30px;
   height: 50px;
   width:100px
 }
 .image-5{
   background: RGB(255,0,0);
   transform: rotate3d(1, -5, 1, -60deg);
   color: white;
   padding: 30px;
   height: 50px;
   width:100px
 }

 p{
   line-height: 2;
   font-size:24px;
 margin-bottom:10px
 }

</style>
</head>
<body>
<div class="demo_container">
   <h2>The rotate3d() function is used to rotate
elements in a three-dimensional space.</h2>

  <div class="row">
```

```
<div class="col">
  <p> Not rotated </p>
  <div class="image-1"> </div>
  <p> Rotate along the x Axis  </p>
  <div class="image-2"> </div>
</div>
<div class="col">
  <p> Rotate along all Axes   </p>
  <div class="image-4"> </div>
    <p> Negative Values </p>
    <div class="image-5"> </div>
</div>
</div>
</div>
</body>
</html>
```

The output of the code is given below:

The rotate3d() function is used to rotate elements in a three-dimensional space.

Not rotated

Rotate along all Axes (rotate3d(1, 5, 1, 60deg)

Rotate along the x Axis (rotate3d(1, 0, 0, 60deg))

Negative Values (rotate3d(1, -5, 1, -60deg))

CSS function (rotate3d ()).

rotateX()

The CSS rotateX() is used to rotate elements around the x-axis in three-dimensional space. The rotateX() is used in 3D-transforms. It's used with the transform property to rotate an element around the x-axis. It can be used in other rotation functions such as rotateY(), rotateZ() to rotate the element around the y and z axes if required.

The syntax of the rotateX() function is as follows:

```
rotateX() = rotateX( <angle> )
```

The angle can be represented with the following unit identifiers:

- deg
- grad
- rad
- turn

rotateY()

The CSS rotateY() is used to rotate elements around the x-axis in three-dimensional space. The rotateY() function is used in 3D-transforms. It is used with the transform property to rotate an element around the x-axis.

The syntax of the rotateY() function is as follows:

```
rotateY() = rotateY( <angle> )
```

The angle can be represented with any of the following identifiers:

- deg
- grad
- rad
- turn

rotateZ()

The CSS rotateZX() function is used to rotate elements around the x-axis in three-dimensional space. The rotateZ() is used in 3D-transforms. It's used with the transform property to rotate an element around the y-axis.

The syntax of the rotateZ() function is as follows:

```
rotateZ() = rotateZ( <angle> )
```

The angle can be represented with any of the following identifiers:

- deg
- grad
- rad
- turn

Example of rotateX(), rotateY(), rotateZ():

```
<html>
<head>
<style>

* {
  padding:0;
  margin:0;
  box-sizing: border-box;
}

.demo_container{
  padding:20px;
  width:900px;
  margin:0 auto;
  justify-content: center;
  align-items: center;
}

.row{
  display: flex;
  flex-wrap: wrap;
  flex-direction: row;
}
.col{
width:50%
}
p{
  height:50px
}
.image-1 {
  background: RGB(255,69,0);
    transform: rotateX(10deg);
    color: white;
    padding: 10px;
    height: 50px;
    width:100px
  }
  .image-2 {
    background: RGB(0,255,0);
    transform: rotateX(-60deg);
```

```css
    color: white;
    height: 50px;
    width:100px;
    padding: 10px;

}

.image-3 {
background: RGB(255,69,0);
  transform: rotateY(10deg);
  color: white;
  padding: 10px;
  height: 50px;
  width:100px
 }
.image-4 {
  background: RGB(0,255,0);
  transform: rotateY(-60deg);
  color: white;
  height: 50px;
  width:100px;
  padding: 10px;

}

.image-5 {
background: RGB(255,69,0);
  transform: rotateZ(10deg);
  color: white;
  padding: 10px;
  height: 50px;
  width:100px
 }
.image-6 {
  background: RGB(0,255,0);
  transform: rotateZ(-60deg);
  color: white;
  height: 50px;
  width:100px;
  padding: 10px;

}
```

```
  p{
    line-height: 2;
    font-size:24px;
  margin-bottom:10px
  }

</style>
</head>
<bo>
<div class="demo_container">
  <h2> The rotateX() function is used to rotate elements
around the x-axis in a three-dimensional space.</h2>
  <div class="row">
    <div class="col">
      <p> rotateX() </p>
      <div class="image-1"> rotateX(10deg) </div>
      <p> Negative Values </p>
      <div class="image-2"> rotateX(-60deg) </div>
    </div>

  </div>
  </div>

<div class="demo_container">
    <h2> The rotateY() function is used to rotate
elements around the y-axis in a three-dimensional
space.</h2>
    <div class="row">
      <div class="col">
        <p> rotateX() </p>
        <div class="image-3"> rotateY(10deg) </div>
        <p> Negative Values </p>
        <div class="image-4"> rotateY(-60deg) </div>
      </div>

    </div>
  </div>

<div class="demo_container">
    <h2> The rotateZ() function is used to rotate
elements around the z-axis in a three-dimensional
space.</h2>
    <div class="row">
      <div class="col">
```

```
        <p>  rotateZ() </p>
        <div class="image-5"> rotateZ(10deg) </div>
        <p> Negative Values </p>
        <div class="image-6"> rotateZ(-60deg) </div>
    </div>

    </div>
  </div>

</body>
</html>
```

The output of the code is given below:

The rotateX() function is used to rotate elements around the x-axis in a three-dimensional space.

rotateX()

Negative Values

The rotateY() function is used to rotate elements around the y-axis in a three-dimensional space.

rotateX()

Negative Values

The rotateZ() function is used to rotate elements around the z-axis in a three-dimensional space.

rotateZ()

Negative Values

CSS function (rotateX(), rotateY (), rotateZ()).

saturate()

The saturate() function is used to adjust the saturation of an image. This function is used with the filter to adjust the saturation levels in an image.

The syntax of the saturate() function is as follows:

```
saturate() = saturate( [ <number> | <percentage> ] )
```

Example:

```
<!DOCTYPE html>
<html>
<head>
<style>

* {
  padding:0;
  margin:0;
  box-sizing: border-box;
}

.demo_container{
  padding:20px;
  width:800px;
  margin:0 auto;
  justify-content: center;
  align-items: center;
}

img{
  width:100%;
  height:250px;
}
.row{
  display: flex;
}
```

```
.col{

width:100%
}
.image-1 {
  filter: saturate(700%);
  }
  .image-2 {
    filter: saturate(2.5);
  }


  p{
    padding-top: 20px;
    font-size:20px
  }
</style>
</head>
<body>
<div class="demo_container">
  <h2> The saturate() function to adjust the
saturation of an image. </h2>
  <div class="row">
    <div class="col">
      <p> Normal Image </p>
      <img src="/images-1.jpg" alt="Sample
image">
      <p> 700% Saturate Image </p>
      <img class="image-1" src="/images-1.jpg"
alt="Sample image">
    </div>
    <div class="col">
      <p> 2.5 Saturate Image </p>
    <img class="image-2" src="/images-1.jpg"
alt="Sample image">
    </div>
  </div>
  </div>
</body>
</html>
```

The output of the code is given below:

The saturate() function to adjust the saturation of an image.

Normal Image   2.5 Saturate Image

700% Saturate Image

CSS function (saturate()).

scale()

The CSS scale() is used to scale elements in two-dimensional space. The scale() scales an element based on the number that provides as an argument. You can scale in the direction of the x-axis, the y-axis, or both. The scale() function works like this:

- scale(sx) or
- scale(sx, sy)
- scale(sx) or
- scale(sx, sy)

The syntax of the scale() function is as follows:

```
scale() = scale( <number> [, <number> ]? )
```

Example:

```
<!DOCTYPE html>
<html>
<head>
<style>

* {
  padding:0;
  margin:0;
  box-sizing: border-box;
}

.demo_container{
  padding:20px;
  width:800px;
  margin:0 auto;
  justify-content: center;
  align-items: center;
}

img{
  width:100%;
  height:250px;
}
.row{
  display: flex;
}
.col{

width:100%
}
.image-1 {
  transform-origin: top left;
    transform: scale(2);
    width:100%;
  height:250px;
  }

  p{
    padding-top: 20px;
    font-size:20px
  }
</style>
</head>
```

```
<body>
<div class="demo_container">
  <h2> The scale() function is used to scale
elements in a two-dimensional space. </h2>
  <div class="row">
    <div class="col-1">
      <p> Normal Image </p>
      <img src="/images-1.jpg" alt="Sample image">
      <p> Scale along both Axes  </p>
      <img class="image-1" src="/images-1.jpg"
alt="Sample image">
    </div>
  </div>
  </div>
</body>
</html>
```

The output of the code is given below:

The scale() function is used to scale elements in a two-dimensional space.

Normal Image

Scale along both Axes

CSS function (scale()).

scale3d()

The scale3d() is used to scale elements in three-dimensional space. The scale3d() scales an element based on the numbers that provide an argument. It specifies a 3D scale operation using the [sx,sy,sz] scaling vector described by three parameters. The scale3d() function is a three-dimensional primitive, with the following derived functions:

scalex(), scaleY(), scaleZ(), and scale().

The syntax of the scale3d() function is as follows:

```
scale3d() = scale3d( <number>, <number>, <number> )
```

Example:

```
<!DOCTYPE html>
<html>
<head>
<style>

* {
  padding:0;
  margin:0;
  box-sizing: border-box;
}

.demo_container{
  padding:20px;
  margin:0 auto;
  justify-content: center;
  align-items: center;
}

.row{
  display: flex;
}
.col{

width:100%
}
.image-1 {
  transform-origin: top left;
    transform: scale3d(3, 3, 1);
```

```
    width:50%;
  height:100px;
  }
  .image-2 {
    transform: scale3d(3, 3, 1);
    width:10%;
  height:100px;
  }

  p{
    padding-top: 20px;
    font-size:20px
  }
</style>
</head>
<body>
<div class="demo_container">
  <h2> The scale3d() function is used to scale
elements in a three-dimensional space. </h2>
  <div class="row">
    <div class="col-1">
      <p> Scale3d() with transform-origin </p>
      <img class="image-1" src="/images-1.jpg"
alt="Sample image">
    </div>

      <p> Scale3d() with transform-origin </p>
      <img class="image-1" src="/images-1.jpg"
alt="Sample image">
  </div>
  </div>
</body>
</html>
```

The output of the code is given below:

CSS function (scale3d()).

scaleX()

The CSS scaleX() is used to scale elements in two-dimensional space along the x-axis. The scaleX() scales an element based on the number/s that provide an argument.

The syntax of the scaleX() function is as follows:

```
scaleX() = scaleX( <number> )
```

Example:

```
<!DOCTYPE html>
<html>
<head>
<style>

* {
  padding:0;
  margin:0;
  box-sizing: border-box;
}

.demo_container{
  width:800px;
  padding:20px;
  margin:0 auto;
  justify-content: center;
  align-items: center;
}

.row{
  display: flex;
  flex-wrap: wrap;
  flex-direction: column;
}
.col{
width:50%
}
```

```
.image-1 {
  width:200px;
  height:200px;
  }
  .image-2 {
    transform-origin: top left;
    transform: scaleX(4);
    width:200px;
    height:200px;
  }
  .image-3{
    transform: scaleX(1);
    width:200px;
  height:200px;
  }
  p{
    padding-top: 20px;
    font-size:20px
  }
</style>
</head>
<body>
<div class="demo_container">
  <h2>The scaleX() function is used to scale
elements in a two-dimensional space along the
x-axis. </h2>
  <div class="row">
    <div class="col-1">
      <p> Normal Image </p>
      <img class="image-1" src="/images-1.jpg"
alt="Sample image">


      <p> scaleX(4) with transform-origin
</p>
      <img class="image-2" src="/images-1.jpg"
alt="Sample image">


      <p> scaleX(1) without transform-origin
</p>
```

```
        <img class="image-3" src="/images-1.jpg"
alt="Sample image">
  </div>
  </div>
</body>
</html>
```

The output of the code is given below:

The scaleX() function is used to scale elements in a two-dimensional space along the x-axis.

Normal Image

scaleX(4) with transform-origin

scaleX(1) without transform-origin

CSS function (scaleX()).

scaleY()

The CSS scaleY() is used to scale elements in two-dimensional space along the x-axis. The scaleY() scales an element based on the number/s that provide an argument. The syntax of the scaleY() function is as follows:

```
scaleY () = scaleY ( <number> )
```

Example:

```
<!DOCTYPE html>
<html>
<head>
<style>

* {
  padding:0;
  margin:0;
  box-sizing: border-box;
}

.demo_container{
  width:800px;
  padding:20px;
  margin:0 auto;
  justify-content: center;
  align-items: center;
}

.row{
  display: flex;
  flex-wrap: wrap;
  flex-direction: row;
}
.col{
width:50%
}
.image-1 {
  width:200px;
  height:200px;
  }
  .image-2 {
    transform-origin: top right;
    transform: ScaleY(2);
```

```
    width:200px;
  height:200px;
  }
  .image-3{
    transform: ScaleY(3);
    width:200px;
    float:right;
  height:200px;
  }
  p{
    padding-top: 20px;
    font-size:20px
  }
</style>
</head>
<body>
<div class="demo_container">
  <h2>The scaleY() function is used to scale
elements in a two-dimensional space along the
y-axis. </h2>
  <div class="row">
    <div class="col">
      <p> Normal Image  </p>
      <img class="image-1" src="/images-1.jpg"
alt="Sample image">

      <p> ScaleY() with  transform-origin  </p>
      <img class="image-1" src="/images-1.jpg"
alt="Sample image">

      <p> ScaleY(4) without transform-origin
</p>
      <img class="image-2" src="/images-1.jpg"
alt="Sample image">

  </div>
  </div>
</body>
</html>
```

The output of the code is given below:

The scaleY() function is used to scale elements in a two-dimensional space along the y-axis.

Normal Image

ScaleY() with transform-origin

ScaleY(4) without transform-origin

CSS function (scaleY()).

scaleZ()

The CSS scaleZ () is used to scale elements in two-dimensional space along the x-axis. The scaleZ() scales an element based on the number/s that provide an argument. The syntax of the scaleZ () function is as follows:

```
scaleZ () = scaleZ ( <number> )
```

Example:

```
<!DOCTYPE html>
<html>
<head>
<style>

* {
  padding:0;
  margin:0;
  box-sizing: border-box;
}

.demo_container{
  width:800px;
  padding:20px;
  margin:0 auto;
  justify-content: center;
  align-items: center;
}

.row{
  display: flex;
  flex-wrap: wrap;
  flex-direction: row;
}
.col{
width:50%
}
.image-1 {
  width:200px;
  height:200px;
  }
  .image-2 {
    transform-origin: top right;
    transform: perspective(250px) scaleZ(2)
rotateX(45deg);
    width:200px;
  height:200px;
  }
  .image-3{
    transform: perspective(250px) scaleZ(2)
rotateX(45deg);
```

```
      width:200px;
      float:right;
   height:200px;
   }
   p{
      padding-top: 20px;
      font-size:20px
   }
</style>
</head>
<body>
<div class="demo_container">
   <h2>The scaleZ() function is used to scale
elements in a two-dimensional space along the
z-axis. </h2>
   <div class="row">
      <div class="col">
         <p> Normal Image </p>
         <img class="image-1" src="/images-1.jpg"
alt="Sample image">

         <p> scaleZ(4) with transform-origin
</p>
         <img class="image-2" src="/images-1.jpg"
alt="Sample image">
         </div>
         <div class="col">
            <p> scaleZ(4) without transform-origin
</p>
            <img class="image-3" src="/images-1.jpg"
alt="Sample image">
         </div>

   </div>
   </div>
</body>
</html>
```

The output of the code is given below:

The scaleZ() function is used to scale elements in a two-dimensional space along the z-axis.

Normal Image

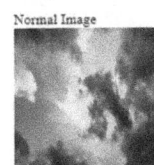

scaleZ(4) without transform-origin

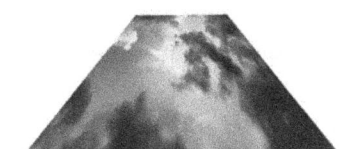

scaleZ(4) with transform-origin

CSS function (scaleZ()).

skew()

The CSS skew() function is used to skew elements in two-dimensional space. The skew() element performs a shear transformation (also known as a shear mapping or a transvection), which displaces each point of an element by a given angle in each direction.

Skewing an element is kind of like taking the points of an element, and pushing or pulling them in different directions, based on a given angle.

The skew() function works like this:

- skew(ax) or

- skew(ax, ay)

Example:

```
<!DOCTYPE html>
<html>
<head>
<style>
```

```
* {
  padding:0;
  margin:0;
  box-sizing: border-box;
}

.demo_container{
  width:800px;
  padding:20px;
  margin:0 auto;
  justify-content: center;
  align-items: center;
}

.row{
  display: flex;
  flex-wrap: wrap;
  flex-direction: row;
}
.col{
width:50%
}
.image-1 {
  width:200px;
  height:200px;
  }
  .image-2 {
    transform-origin: top left;
    transform: skew(10deg, 0);
    width:200px;
  height:200px;
  }
  .image-3{
    transform-origin: top left;
    transform: skew(0, 10deg);
      width:200px;
  height:200px;
  }
  .image-3{
    transform-origin: top left;
    transform: skew(0, 10deg);
      width:200px;
  height:200px;
  }
```

```
   .image-4{
     transform-origin: top left;
   transform: skew(10deg, 10deg);
     width:200px;
 height:200px;
 }

 .image-5{
   transform-origin: bottom left;
   transform: skew(-10deg, -10deg);
        width:200px;
 height:200px;
 }

   .image-6{
   transform: skew(-10deg, -10deg);
        width:200px;
 height:200px;
 }

 p{
   padding-top: 20px;
   font-size:20px
 }
</style>
</head>
<body>
<div class="demo_container">
  <h2> The skew() function is used to skew
elements in a two-dimensional space.  </h2>
  <div class="row">
    <div class="col">
      <p> Normal Image</p>
      <img class="image-1" src="/images-1.jpg"
alt="Sample image">

      <p> Skew along the x-axis  </p>
      <img class="image-2" src="/images-1.jpg"
alt="Sample image">

      <p> Skew along the y-axis </p>
      <img class="image-3" src="/images-1.jpg"
alt="Sample image">
      </div>
```

```
    <div class="col">
    <p> Skew along Both Axes  </p>
    <img class="image-4" src="/images-1.jpg"
alt="Sample image">

    <p> Negative Values  </p>
    <img class="image-5" src="/images-1.jpg"
alt="Sample image">

    <p> Removing transform-origin </p>
    <img class="image-6" src="/images-1.jpg"
alt="Sample image">

    </div>
  </div>
  </div>
</body>
</html>
```

The output of the code is given below:

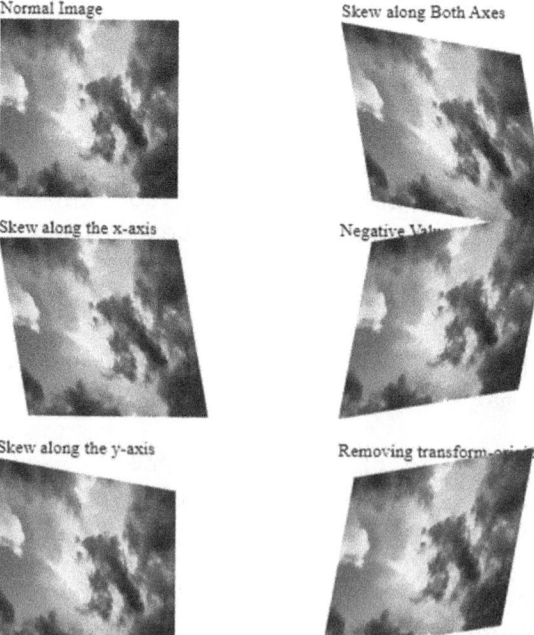

CSS function (skew()).

skewX()

The skewX() element performs a shear transformation (also known as a shear mapping or a transvection), which displaces each point of an element by a given angle along the x-axis.

The skewX() function works like this:

```
skewX(a)
```

The skewX() function accepts one argument which specifies the angle of the skew for the x-axis. It can be any valid angle value.

Example:

```
<!DOCTYPE html>
<html>
<head>
<style>

* {
  padding:0;
  margin:0;
  box-sizing: border-box;
}

.demo_container{
  width:800px;
  padding:20px;
  margin:0 auto;
  justify-content: center;
  align-items: center;
}

.row{
  display: flex;
  flex-wrap: wrap;
  flex-direction: row;
}
.col{
width:50%
}
```

```css
.image-1 {
  width:200px;
  height:200px;
  }
  .image-2 {
    transform-origin: top left;
    transform: skewX(10deg);
    width:200px;
  height:200px;
  }
  .image-3 {
    transform-origin: top left;
    transform: skewX(-10deg);
    width:200px;
  height:200px;
  }

  p{
    padding-top: 20px;
    font-size:20px
  }
</style>
</head>
<body>
<div class="demo_container">
  <h2> The skewX() function is used to skew
elements in a two-dimensional space along the
x-axis.  </h2>
  <div class="row">
    <div class="col">
      <p> Normal Image </p>
      <img class="image-1" src="/images-1.jpg"
alt="Sample image">

      <p> Skew with transform-origin </p>
      <img class="image-2" src="/images-1.jpg"
alt="Sample image">
      </div>
```

```
        <div class="col">
        <p> Negative Values        </p>
        <img class="image-3" src="/images-1.jpg"
    alt="Sample image">


        </div>
    </div>
    </div>
</body>
</html>
```

The output of the code is given below:

The skewX() function is used to skew elements in a two-dimensional space along the x-axis.

Normal Image

Negative Values

Skew with transform-origin

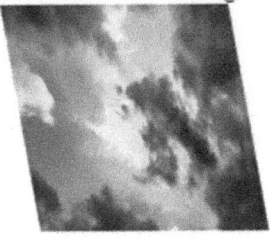

CSS function (skewX()).

CHAPTER SUMMARY

In this chapter, we saw various built-in methods like attr(), RGB(), rgba(), and so on. These functions have their own unique functionality and properties. So next chapter is about CSS plugins.

CHAPTER 5

CSS Plugins

IN THIS CHAPTER

- ➤ Introduction
- ➤ Various CSS Plugins Based on PostCSS
- ➤ Autoprefixer
- ➤ CSSnext
- ➤ CSS MarqueeMenu Plugin

In the last chapter, we studied about CSS functions. In this chapter, we will discuss some plugins that are based on CSS and also work with JavaScript or any other programming language. So PostCSS plays a major role in it. All other plugins are defined in PostCSS. PostCSS is a versatile tool that allows to transform CSS styles using JavaScript plugins. Its flexibility lies in the way it is built. The core of PostCSS is a Node.js module that you can install using npm, it has over 200 plugins that you can choose to use in your project. PostCSS is a framework for developing various CSS tools. It can be used to develop template languages like Sass and LESS. PostCSS consists of a CSS parser that generates an abstract syntax tree.

We also have pre- and post-processing scripts that run before an item is saved. The difference between the two is that the pre-processing scripts run before the value checking and validation rules are completed, and the postprocessing scripts run after these processes.

DOI: 10.1201/9781003358060-5

PostCSS is neither preprocessor nor postprocessor, as various PostCSS plugins can fall into one or both of these categories; it's entirely up to you what you make of it. With PostCSS, you don't have to learn another syntax like Sass or LESS; you can start using it immediately.

PostCSS will take your existing CSS file and convert it to JavaScript readable data, then the JavaScript plugins will make the modifications and PostCSS will return the modified version of the original file.

VARIOUS CSS PLUGINS PostCSS BASED

There are various plugins available in the market as given below:

1. Autoprefixer

2. CSSnext

3. CSSNano

4. PreCSS

5. StyleLint

6. PostCSS Assets

7. Font Magician

8. Lost Grid

9. PostCSS

10. PostCSS-modules

AUTOPREFIXER

Autoprefixer uses the data based on the latest browser popularity that supports to apply prefixes for you. You can try the interactive demo of Autoprefixer.

You can write any CSS rules without vendor prefixes as given below:

```
.container {
    box-shadow: 0 0 10px #cecece;
    transition: all .5s;
}
```

p{

```
color: red;
}

.section a{
display: flex;
}
```

Write Pure CSS

Working with Autoprefixer is simple, you need to forget about vendor pre-fixes and write normal CSS according to the latest W3C specs. You do not need a special language (like Sass, Less) or remember where you must use mixins.

Autoprefixer supports selectors (such as :fullscreen & ::selection), unit function (calc()), at-rules (@support & @keyframes) and with properties.

It is a postprocessor, you can use it with preprocessors like Sass, Stylus, or LESS.

Just write normal CSS according to the latest W3C specs and Autoprefixer will produce the code for old browsers. Autoprefixer changes CSS indentation to create a nice visual cascade of prefixes if the CSS is uncompressed.

```
a {
    -webkit-box-sizing: border-box;
       -moz-box-sizing: border-box;
            box-sizing: border-box;
}
```

The output with Autoprefixer will be like this (you can check the Autoprefixers code online using, https://tools.webdevpuneet.com/css-autoprefixer/):

```
/*
 * Prefixed by:
 * PostCSS: v7.0.29,
 * Autoprefixer: v9.7.6
 * Browsers: last 4 version
 */

.container {
    -webkit-box-shadow: 0 0 10px #cecece;
            box-shadow: 0 0 10px #cecece;
    -webkit-transition: all .5s;
```

```
    -o-transition: all .5s;
    transition: all .5s;
  }

p{
color: red;
}

.section a{
display: -webkit-box;
display: -ms-flexbox;
display: flex;
}
```

The thing is that some CSS properties will display differently in different browsers, and you want every site visitor to have a predictable and accurate experience. For older, more established properties, such as a border or border, there is nothing to worry about; they have been standardized and work across browsers. But newer features like grid or flexbox are not so simple.

The vendors that make your browsers, like Google and Mozilla, don't want to wait for cool new things to be standardized before implementing them. Instead, they create their own version and add a prefix that differentiates it from other browsers' implementations. The new property will be standardized and work the same across all browsers, eliminating the need for prefixes. Until then, you must use vendor prefixes in your code.

"Autoprefixer" is a plugin that can save us from the monotony of -webkit- and -moz-. It does exactly sounds like: it automatically adds prefixes to your CSS. All you have to do is supply it using a CSS sheet that will read it and add vendor prefixes if necessary.

Installing Autoprefixer in Your Project Folder

Open a terminal in VS code and use npm to install it, well as postcss and its companion postcss-cli, the command-line tool you'll use to run Autoprefixer. The process is the same as installing Sass, only you install three packages at once. You can install so many at once you'd like, each separated by a space as

```
$ npm install autoprefixer postcss postcss-cli -g
```

Once npm has downloaded and installed the packages, need to go back into package.json and add a script for npm to run, just like when you installed Sass. Add a new script named "prefix" after the "sass"'s script:

```
{
  "name": "abc",
  "version": "1.0.0",
  "description": abc",
  "main": "index.js",
  "scripts": {
    "sass": "sass ./sass/main.scss:./public/css/style.
css -w --style compressed",
    "prefix":
  },
  "author": "",
  "license": "ISC",
}
```

And inside the script need to tell npm to use the new postcss package you just installed and also where to find your compiled CSS file:

```
{
  "name": "abc",
  "version": "1.0.0",
  "description": "abc",
  "main": "index.js",
  "scripts": {
    "sass": "sass ./sass/main.scss:./public/css/style.
css -w --style compressed",
    "prefix": "postcss ./public/css/style.css"
  },
  "author": "",
  "license": "ISC",
}
```

After that you need to instruct npm which package to use and where to find the CSS file, to use Autoprefixer by implementing – use flag followed by Autoprefixer as given below:

```
{
  "name": "abc",
  "version": "1.0.0",
```

```
  "description": "abc",
  "main": "index.js",
  "scripts": {
    "sass": "sass ./sass/main.scss:./public/css/style.
css -w --style compressed",
    "prefix": "postcss ./public/css/style.css --use
autoprefixer"
  },
  "author": "",
  "license": "ISC",
}
```

And then, finally, you need to instruct it where to put new, prefixed, CSS sheet as given below:

```
{
  "name": "abc",
  "version": "1.0.0",
  "description": "abc",
  "main": "index.js",
  "scripts": {
    "sass": "sass ./sass/main.scss:./public/css/style.
css -w --style compressed",
    "prefix": "postcss ./public/css/style.css --use
autoprefixer -d ./public/css/prefixed/"
  },
  "author": "",
  "license": "ISC",
}
```

Now your prefixing script is complete.
Add a new key named browserslist as given below:

```
{
  "browserslist":
}
```

And then give browserslist a value of last four versions :

```
{
  "name": "joeblow",
  "version": "1.0.0",
```

```
    "description": "Joe Blow's web portfolio",
    "main": "index.js",
    "scripts": {
      "sass": "sass ./sass/main.scss:./public/css/style.
css -w --style compressed",
      "prefix": "postcss ./public/css/style.css --use
autoprefixer -d ./public/css/prefixed/"
    },
    "author": "",
    "license": "ISC",
    "browserslist": "last 4 versions"
}
```

Now you are ready to autoprefix CSS! Let's move into the terminal and run the prefix script:

```
$ npm run prefix
```

Take a look at the new, prefixed, CSS in your CSS file, a new block will be added:

```
.section {
  background: #15DEA5;
  height: 10rem;
  display: -webkit-box;
  display: -ms-flexbox;
  display: flex;
  -webkit-box-align: center;
      -ms-flex-align: center;
          align-items: center;
  -ms-flex-wrap: wrap;
      flex-wrap: wrap;
  width: 100%;
}
```

Now our website will display properly and uniformly across all compatible browsers.

When Does It Run?

Autoprefixer is the very last thing that runs when processing your CSS. So if you use Sass, LESS, or Stylus, that will process first, then Autoprefixer. You are still free to use a preprocessor add-on (e.g., Compass, LESSHat, etc.)

in which to use for prefixing if you'd like, but I'd argue that Autoprefixer is easier, does a better job, and you don't need both.

What is the Difference Between Autoprefixer and -Prefix-Free?

-Prefix-free is a client-side JavaScript library. It is included on your page as an additional request. It runs, finds all the CSS it can find, figures out what prefixes may be needed, and injects that new CSS onto the page after the CSS it finds.

- Downsides: extra request, flash-of-un-prefixed-styles, not cached, a few styles it can't do (e.g., filter).

- Upsides: smaller CSS, only injects prefixes current browser needs.

Autoprefixer processes your CSS server-side. At least, that's how we're doing it because that seems to be the canonical source for the project.

- Downsides: larger CSS (but not as big as "just prefixing everything").

- Upsides: prefixes perfectly to what you support, final CSS is cacheable, handles tricky fallbacks (e.g., flexbox).

CSSnext

PostCSS is a new tool that makes easy to develop JavaScript plugins that transform styles. This opens up a new world of possibilities for new plugins that make working with CSS easier and easier. The post introduces two popular PostCSS plugins: cssnext and cssnano.

cssnext lets you use the future of CSS today. You can use features that are not supported in all browsers, such as CSS variables and CSS color functions. cssnext transforms your styles to work in all browsers. In other words, cssnext allows to write styles with real CSS syntax instead of another preprocessor syntax. cssnext will automatically add vendor prefixes to your styles, so you dont have to use the prefixes yourself when writing CSS.

To give you an example, let's say you have the following CSS styles:

```
:root {
  --text: pink;
  --bg-color: #FFEC31;
  --flex-center: {
```

```
    display: flex;
    margin: auto;
  }
}

.box {
  background-color: var(--bg-color);
  color: color(hotpink whiteness(25%));
  @apply(--flex-center);
}

.warn {
  @apply(--flex-center);
}
```

cssnext will transform the styles into this:

```
.box {
  background-color: #F9EC31;
  color: rgb(255, 64, 159);
  display: -webkit-box;
  display: -ms-flexbox;
  display: flex;
  margin: auto;
}

.warn {
  display: -webkit-box;
  display: -ms-flexbox;
  display: flex;
  margin: auto;
}
```

This plugin is the Babel of CSS, it allows you to use modern CSS features and at the same time takes care of translating them into CSS that is more digestible for older browsers:

- adds prefixes using Autoprefixer (so if you use this you don't need to use Autoprefixer directly).

- allows you to use CSS variables.

- allows you to use nesting, like in Sass.

CSSnano

cssnano minifies and compresses your CSS. It removes whitespace, eliminates duplicate rules, outdated vendor prefixes, comments, and performs a lot of other optimizations. Both cssnext and cssnano can be configured to work according to specific needs.

Installing PostCSS, the PostCSS-CLI, cssnext, and cssnano

Using npm:

```
$ npm install --save-dev postcss postcss-cli postcss-cssnext cssnano
```

Or through Yarn command:

```
$ yarn add postcss postcss-cli postcss-cssnext cssnano --dev
```

Using the PostCSS-CLI

You use the PostCSS command line (cmd) interface by giving the input, output files, and PostCSS plugin(s) to use. It specifies the plugins with the – use flag, the output file with the – output flag, and the input file is provided last without any flags using the below command:

```
$ postcss --use postcss-cssnext --use cssnano --output styles-out.css styles.css
```

You can use the PostCSS CLI in watch mode, to listen for changes to the input file using the below command:

```
$ postcss --use postcss-cssnext --use cssnano --output styles-out.css styles.css --watch
```

You can specify more fine-grained configuration options in a json configuration file to specify the config file with the – config flag:

```
$ postcss --config postcss-config.json
```

Your config file will look like this:

```
{
    "use": ["postcss-cssnext", "cssnano"],
```

```
        "input": "styles.css",
        "output": "styles-out.css"
}
```

npm Script

To make your workflow run smooth, simply set up a postcss script in your project's package.json file:

```
"scripts": {
    "postcss": "postcss --use postcss-cssnext --use
cssnano --output styles-out.css styles.css"
}
```

Now all you have to do is run the following command:

```
$ npm run postcss
```

PreCSS

It will let you use Sass-like markup and staged CSS features in CSS. This is the given example:

```
$blue: #056ef0;
$column: 200px;

.menu {
    width: calc(4 * $column);
}

.menu_link {
    background: $blue;
    width: $column;
}
```

The result will be like this:

```
.menu {
    width: calc(4 * 200px);
}

.menu_link {
    background: #056ef0;
    width: 200px;
}
```

It combines Sass-like syntactical suga like variables, conditionals, and iterators, with emerging CSS features, like logical and custom properties, media query ranges, and image sets.

Usage

You can add PreCSS to your build tool:

```
$ npm install precss --save-dev
```

Other Various Plugins

It has powered by the following plugins (in this order):

- postcss-extend-rule: You can add this PostCSS Extend Rule to your project:

```
$ npm install postcss postcss-extend-rule
--save-dev
```

Use PostCSS Extend Rule to process your CSS:

```
const postcssExRule = require('postcss-
extend-rule');

postcssExRule.process(CSS_CODE /*, processOptions,
pluginOptions */);
```

There are various Options available in this plugins.

The name option determines the at-rule name being used to extend selectors. By default, this name is extend, meaning @extend rules are parsed.

```
postcssExtend({ name: 'postcss-extend' })
```

If the name option were changed to, say, postcss-extend, then only @postcss-extend at-rules would be parsed.

```
main {
    @postcss-extend.some-rule;
}
```

- postcss-advanced-variables: Its variables let you use Sass-like variables, conditionals, and iterators in CSS.

  You can add PostCSS advanced variables to your build tool:

```
$ npm install postcss-advanced-variables
--save-dev
```

You can add PostCSS to your build tool:

```
$ npm install postcss --save-dev
```

Use PostCSS advanced variables as a plugin:

```
postcss([

  require('postcss-advanced-variables')(/* options
*/)

]).process(YOUR_CSS);
```

- postcss-preset-env: It lets us to convert modern CSS into something that most browsers can understand, determining the polyfills need based on targeted browsers or runtime environments.

  ```
  $ npm install postcss-preset-env
  ```

  You can add PostCSS Preset Env into your project:

  ```
  $ npm install postcss-preset-env --save-dev (run
  this command in the terminal)
  ```

  Now use PostCSS Preset Env to process your CSS as given below:

  ```
  const postCSSPresetEnv = require('postcss-
  preset-env');

  postCSSPresetEnv.process(YOUR_CSS /*,
  processOptions, pluginOptions */);
  ```

Various Options

1. **stage:** This option determines which CSS features to polyfill, based on the stability in the process of becoming implemented web standards.

   ```
   postcssPresetEnv({ stage: 0 })
   ```

   This stage can be 0 (experimental) through 4 (stable), or false. By setting stage to false will disable all polyfill. This would be useful if you intended to use the features option exclusively.

   Without any configuration options, PostCSS allows Preset Env Stage.

2. **features:** The features enable or disable specific polyfills by ID. By passing true to a specific feature ID will enable its polyfill, during passing false will disable it.

There are various id available in this plugin as given below:

1. postcssAttributeCaseInsensitive

2. postcssBlankPseudo

3. postcssColorFunctionalNotation

4. postcssEnvFunction

5. postcssLogical

6. postcssReplaceOverflowWrap

7. postcssSelectorNot, and so on

You can use these IDs just by importing them in the code.

3. **insertBefore / insertAfter:** The insertBefore and insertAfter keys allow to insert other PostCSS plugins into the chain. This is useful if you are using sugary PostCSS plugins that execute before or after particular polyfills. Both insertBefore and insertAfter support chaining with one or more multiple plugins.

4. **autoprefixer:** It also includes autoprefixer and browsers option will be passed to it automatically. It specifying the autoprefixer option enables passing additional options into autoprefixer.

```
postcssPresetEnv({
  autoprefixer: { grid: true }
})
```

• postcss at-root: The @at-root causes one or more rules to be emitted at the root of a document, rather than nested beneath their parent selectors:

```
.section {
  ...
  @at-root{
.    div {...}
  }
}
```

The result will be as given below:

```
.child { ... }
.parent { ... }
```

How to use it in code:

```
postcss([ require('postcss-atroot')() ])
```

- postcss-property-lookup:PostCSS plugin that allows referencing property values without a variable, similar to Stylus.

```
.section {
  margin-left: 20px;
  margin-right: @margin-left;
  color: red;
  background: @color url('test.png');
  line-height: 1.5;
  font-size: @(line-height)em;
}
.section {
  margin-left: 20px;
  margin-right: 20px;
  color: red;
  background: red url('test.png');
  line-height: 1.5;
  font-size: 1.5em;
}
```

How to use in code:

```
postcss([ require('postcss-property-lookup') ])
```

- postcss-nested

Install Plugin

- Step 1:

```
$ npm install --save-dev postcss postcss-nested
```

- Step 2: Check your existing project for PostCSS config: postcss.config.js in the project root, "postcss" section in package.json or postcss in bundle config.

If you don't use PostCSS, add it according to official docs and set this plugin in settings.

- Step 3: Add the plugin to plugins list:

```
module.exports = {
  plugins: [
+   require('postcss-nested'),
    require('autoprefixer')
  ]
}
```

There are various options available:

1. bubble: By default, plugin will bubble only the @media and @ supports at-rules. You can add custom at-rules to list by bubble option:

```
postcss([ require('postcss-nested')({ bubble:
['phone'] }) ])
```

2. unwrap: By default, plugin will unwrap the @font-face, @keyframes, and @document at-rules. You can add custom at-rules to the list by unwrap option:

```
postcss([ require('postcss-nested')({ unwrap:
['phone'] }) ])
```

Example:

```
/* Enter your CSS code */
a {
  color: white;
  @phone {
    color: black;
  }
}
/* The output will be like this */
a {
  color: white;
}
@phone {
  color: black;
}
```

preserveEmpty

By default, plugin will strip out any empty selector generated by intermediate nesting levels. You can set preserveEmpty to true to preserve them.

```
.a {
    .b {
        color: black;
    }
}
```

It will be compiled as:

```
.a { }
.a .b {
    color: black;
}
```

You can install postcss-nested using the command.

```
$ npm i postcss-nested
```

STYLELINT

It is a mighty, modern CSS linter that helps to enforce consistency and
avoid errors in your stylesheets.

Installing Stylelint

First, ensure you have stylelint installed. It is an npm package which can be
installed by running the following command:

```
$ npm install -g stylelint
```

Usage

After you have stylelint installed, you will want to create a .stylelintrc.json
file. This is where will configure all the lint rules want stylelint to check
for. Then take a look at the documentation rules page for details of all the
available rules.

Once have a . stylelintrc.json file set up with lint rules, you can run
stylelint on SCSS files with the following command. This assumes that you
are in the directory where .scss files are located. You can update the path
accordingly.

It will look for a .stylelintrc.json file in the present working directory. If any of your config file is not in the present working directory, you can pass it along when run stylelint with the config option.

```
stylelint "**/*.scss" --config foo/bar/.stylelintrc.json
```

If you want to use stylelint with npm-scripts, use the below command:

```
$ npm install stylelint -D
```

PostCSS ASSETS

A PostCSS Assets is an asset manager for CSS. It isolates stylesheets from environmental changes, gets image sizes and inlines files.
Installation:

```
$ npm install postcss-assets --save-dev
```

Gulp (gulp-postcss)
PostCSS gulp plugin to pipe CSS through several plugins, but parse CSS only once.
Installation of gulp-postcss:

```
$ npm install --save-dev postcss gulp-postcss
```

Basic
The configuration is loaded by own from postcss.config.js as described here, so you don't have to specify any options.

```
var postcss = require('gulp-postcss');
var gulp = require('gulp');

gulp.task('css', function () {
    return gulp.src('./src/*.css')
        .pipe(postcss())
        .pipe(gulp.dest('./dest'));
});
```

Passing plugins directly using gulp:

```
var postcss = require('gulp-postcss');
var gulp = require('gulp');
```

```
var autoprefixer = require('autoprefixer');
var cssnano = require('cssnano');

gulp.task('css', function () {
    var plugins = [
        autoprefixer({browsers: ['last 1 version']}),
        cssnano()
    ];
    return gulp.src('./src/*.css')
        .pipe(postcss(plugins))
        .pipe(gulp.dest('./dest'));
});
```

Grunt

Now we will explain how to create a Gruntfile and install and use Grunt plugins. Once you are familiar with that process, you can install this plugin with this command:

```
$ npm install grunt-postcss --save-dev
```

Once the plugin has been installed, then may enable inside Gruntfile with this line of JavaScript:

```
grunt.loadNpmTasks('grunt-postcss');
```

How to use the plugins in the code:

```
$ npm install grunt-postcss pixrem autoprefixer
cssnano
grunt.initConfig({
  postcss: {
    options: {
      map: true, // inline sourcemaps

      // or
      map: {
          inline: false, // save all sourcemaps as
separate files...
          annotation: 'dist/css/maps/' // ...to the
specified directory
      },
```

```
     processors: [
        require('pixrem')(), // add fallbacks for rem
units
        require('autoprefixer')({browsers: 'last 2
versions'}), // add vendor prefixes
        require('cssnano')() // minify the result
     ]
   },
   dist: {
     src: 'css/*.css'
   }
  }
});
```

FONT MAGICIAN

Font Magician is a PostCSS plugin that magically generates all of your @
font-face rules. Never write a @font-face rule again.

Here is an example of font magician:

Just use the font and font-family properties as if they were magic.

```
body {
   font-family: "Alice";
}
```

```
The output will be,

@font-face {
   font-family: "Alice";
   font-style: normal;
   font-weight: 400;
   src: local("Alice"), local("Alice-Regular"),
       url("http://fonts.gstatic.com/s/alice/v7/
sZyKh5NKrCk1xkCk_F1S8A.eot?#") format("eot"),
       url("http://fonts.gstatic.com/s/alice/v7/
l5RFQT5MQiajQkFxjDLySg.woff2") format("woff2"),
       url("http://fonts.gstatic.com/s/alice/v7/_
H4kMcdhHr0B8RDaQcqpTA.woff")  format("woff"),
       url("http://fonts.gstatic.com/s/alice/v7/
acf9XsUhgp1k2j79ATk2cw.ttf")   format("truetype")
}
```

```
body {
  font-family: "Alice";
}
```

Installation: You can add Font Magician to your build tool:

```
$ npm install postcss postcss-font-magician --save-dev
```

or

```
yarn add postcss postcss-font-magician --dev
```

LOST GRID

LostGrid is a powerful grid system built in PostCSS that works with any preprocessor and even vanilla CSS.

Lost Grid is a great PostCSS plugin that provides an impressive CSS grid system that not only works with plain CSS but with preprocessor languages (Sass, LESS, Stylus). It uses the calc() function to create beautiful grids that can be easily used without spending much time with customization.

You can do installation by using the below libraries:

- Gulp: First, you have to install NodeJs then install gulp globally using the below command:

  ```
  $ npm install --global gulp
  ```

  You can also install dependencies:

  ```
  $ npm install --save-dev gulp gulp-postcss gulp-
  sourcemaps autoprefixer lost
  ```

  Now create a gulpfile.js with the following code:

  ```
  var gulp = require('gulp'),
      postcss = require('gulp-postcss'),
      sourcemaps = require('gulp-sourcemaps'),
      autoprefixer = require('autoprefixer'),
      lost = require('lost');

  var paths = {
    cssSource: 'src/css/',
    cssDestination: 'dist/css/'
  };
  ```

```
gulp.task('styles', function() {
  return gulp.src(paths.cssSource + '**/*.css')
    .pipe(sourcemaps.init())
    .pipe(postcss([
      lost(),
      autoprefixer()
    ]))
    .pipe(sourcemaps.write('./'))
    .pipe(gulp.dest(paths.cssDestination));
});

gulp.watch(paths.cssSource + '**/*.css',
['styles']);

gulp.task('default', ['styles']);
```

At last run gulp.

- Grunt: First, you have to install NodeJs then install Grunt using this command:

```
$ npm install --global grunt-cli
```

Install dev dependencies:

```
$ npm install --save-dev grunt grunt-postcss
grunt-autoprefixer grunt-contrib-watch lost
```

Create a Gruntfile.js with the following code:

```
module.exports = function(grunt) {
  grunt.initConfig({
    postcss: {
      options: {
        map: true,
        processors: [
          require('lost')
        ]
      },
      dist: {
        src: 'src/css/style.css',
        dest: 'dist/css/style.css'
      }
    },

    autoprefixer: {
```

```
      single_file: {
        src: 'dist/css/style.css',
        dest: 'dist/css/style.css'
      }
    },

    watch: {
      files: ['src/css/style.css'],
      tasks: ['postcss', 'autoprefixer']
    }

  });

  grunt.loadNpmTasks('grunt-postcss');
  grunt.loadNpmTasks('grunt-autoprefixer');
  grunt.loadNpmTasks('grunt-contrib-watch');
  grunt.registerTask('default', ['watch']);
};
```

- Brunch: First, you install NodeJS then install Brunch using this command:

```
$ npm install -g brunch
```

Create a new Brunch project brunch new
Install PostCSS:

```
$ npm install --save postcss-brunch
```

Install Autoprefixer:

```
$ npm install --save autoprefixer
```

Install Lost:

```
$ npm install --save lost
```

Update brunch-config.coffee
This will be file of export given below:

```
exports.config =
  files:
    javascripts:
      joinTo: 'app.js'
    stylesheets:
  joinTo: 'app.css'
templates:
  joinTo: 'app.js'
```

```
plugins:
  postcss:
    processors: [
      require('autoprefixer')(),
      require('lost')
    ]
```

Now run the brunch using this command:

```
brunch watch
```

There are also various options to install this plugin:

- Stylus

- JavaScript

- Webpack

- Meteor

PostCSS

PostCSS plugin generates spritesheets from your stylesheets.

It is a software development tool that uses JavaScript-based plugins to automate routine CSS operations. It was previously designed by Andrey Sitnik with the idea carrying its origin in his front-end work for Evil Martians.

It is a framework to develop CSS tools. It can also be used to develop a template language such as Sass and LESS.

The PostCSS has core components which are given as below:

- CSS parser generates an abstract syntax tree.

- It is a set of classes that comprise the tree.

- CSS generator generates a line for the object tree.

- Code map generator for the changes made.

Features

Plugins are small programs working with a tree of objects. After the core transforms the CSS string into a tree of objects, the plugins parse and modify the tree. Then PostCSS will generate a new CSS string for the tree changed by the plugin. PostCSS & its plugins are written in JavaScript and installed through npm, which offers an API for various low-level JavaScript operations.

There are official tools that allow you to use PostCSS with build systems such as Webpack, Gulp, and Grunt. A console interface is also available. Browserify or Webpack can use to open PostCSS in a browser. PostCSS plugins perform a variety of CSS processing tasks, from parsing and sorting properties to minification.

A full list of plugins can be found at postcss.parts with some examples listed below:

- Autoprefixer for adding and deleting browser prefixes. CSS modules for isolating CSS selectors and organizing code. It comes as part of Webpack.

- stylelint for analyzing CSS code errors and checking style consistency.

- stylefmt fixes CSS code according to stylelint settings.

- PreCSS to perform some Sass/Less preprocessing functionality.

- postcss-preset-env to emulate features from draft CSS specifications.

- cssnano to reduce CSS size by removing whitespace and rewriting code.

- RTLCSS to change the CSS code so that the design is suitable for right-to-left typing.

- postcss-assets, postcss-inline-svg, and postcss-sprites for working with graphics.

We discuss some of the plugins in brief.

POSTCSS-MODULES (CSS MODULES)

A CSS Module is a CSS file in which all class and animation names are scoped locally by default. All URLs (url(...)) and @imports are in module request format (./xxx and . ./xxx means relative, xxx and xxx/yyy means in modules folder, i. e. in node_modules).

CSS Modules compile to a low-level interchange format called ICSS or Interoperable CSS but are written like normal CSS files:

```
/* style.css */
.className {
  color: green;
}
```

When importing the CSS Module from a JS Module, it exports an object with all mappings from local names to global names.

```
import styles from "./style.css";
// import { className } from "./style.css";
element.innerHTML = '<div class="' + styles.className
+ '">';
```

Naming (localsConvention)

For local class names camelCase naming is recommended:

```
Type: String | (originalClassName: string,
generatedClassName: string, inputFile: string) =>
className: string Default: null
```

Style of exported classnames, the keys in your json.

1. camelCase' - {String}: The class names should be camelized.

2. ' camelCaseOnly' - {String}: The class names will be camelized, and the class name will be removed from the locals.

3. 'dashes' - {String}: It is only dashes in class names that will be camelized.

4. 'dashesOnly'- {String}- It is dashes in class names that will be camelized.

Features

- Modular and reusable CSS.

- No more conflicts.

- Explicit dependencies.

- No global scope.

CSS MarqueeMenu PLUGIN

It is a type of animation effect used in developing web pages for getting attractive text or image scrolling in vertical or horizontal directions.

Note: You can download the CSS MarqueeMenu plugin in the working folder and include the required files in the head section of the HTML code.

Example:

```
<!DOCTYPE html>
<html lang="en" class="no-js">

<head>
  <meta charset="UTF-8" />
  <meta name="viewport" content="width=device-
width, initial-scale=1">
  <title>CSS-only Marquee Menu Effect | Codrops
</title>
  <meta name="description" content="A menu with a
css-only marquee hover effect" />
  <meta name="keywords" content="marquee, css,
animation, loop, infinite, hover, menu,
navigation" />
  <meta name="author" content="Codrops" />
  <link rel="shortcut icon" href="favicon.ico">
  <link rel="stylesheet" href="https://use.
typekit.net/zhq0vyf.css">
  <link rel="stylesheet" type="text/css"
href="style.css" />

</head>

<body class="demo-1">
  <main>

    <nav class="menu">
      <div class="menu__item">
        <a class="menu__item-link"> Stratos (Font
Family) </a>
        <div class="marquee">
          <div class="marquee__inner"
aria-hidden="true">
            <span>Stratos </span>
            <span> Stratos </span>
          </div>
        </div>
      </div>
    </nav>
  </main>
</body>

</html>
```

Style.css

```css
*,
*::after,
*::before {
    box-sizing: border-box;
}

:root {
    font-size: 15px;
}

body {
    margin: 0;
    --color-text: #111;
    --color-bg: #f8ecde;
    --color-link: #b19e7f;
    --color-link-hover: #000;
    color: var(--color-text);
    background-color: var(--color-bg);
    -webkit-font-smoothing: antialiased;
    -moz-osx-font-smoothing: grayscale;
    font-family: stratos, sans-serif;
}

a {
    text-decoration: none;
    color: var(--color-link);
    outline: none;
}

a:hover,
a:focus {
    color: var(--color-link-hover);
    outline: none;
}

.menu {
    -webkit-touch-callout: none;
    -webkit-user-select: none;
    -moz-user-select: none;
    -ms-user-select: none;
    user-select: none;
```

```
    padding: 10vh 0 25vh;
    --marquee-width: 100vw;
    --offset: 20vw;
    --move-initial: calc(-25% + var(--offset));
    --move-final: calc(-50% + var(--offset));
    --item-font-size: 10vw;
    counter-reset: menu;
}

.menu__item {
    cursor: default;
    position: relative;
    padding: 0 5vw;
}

.menu__item-link {
    display: inline-block;
    cursor: pointer;
    position: relative;
    -webkit-text-stroke: 1.5px #000;
    -webkit-text-fill-color: transparent;
    color: transparent;
    transition: opacity 0.4s;
}

.menu__item-link::before {
    all: initial;
    font-family: sofia-pro, sans-serif;
    counter-increment: menu;
    content:  counter(menu);
    position: absolute;
    bottom: 60%;
    left: 0;
    pointer-events: none;
}

.menu__item-link:hover {
    transition-duration: 0.1s;
    opacity: 0;
}


.menu__item-link:hover +. menu__item-img {
    opacity: 1;
```

```
    transform: translate3d(calc(-100% - 6vw),-30%,0)
rotate3d(0,0,1,4deg);
    transition: all 0.4s;
}

.marquee {
    position: absolute;
    top: 0;
    left: 0;
    width: var(--marquee-width);
    overflow: hidden;
    pointer-events: none;
    mix-blend-mode: color-burn;
}

.marquee__inner {
    display: flex;
    position: relative;
    transform: translate3d(var(--move-initial), 0, 0);
    animation: marquee 5s linear infinite;
    animation-play-state: paused;
    opacity: 0;
    transition: opacity 0.1s;
}

.menu__item-link:hover ~. marquee .marquee__inner {
    animation-play-state: running;
    opacity: 1;
    transition-duration: 0.7s;
}

.marquee span {
    text-align: center;
}

.menu__item-link,
.marquee span {
    white-space: nowrap;
    font-size: var(--item-font-size);
    padding: 0 1vw;
    font-weight: 900;
    line-height: 1.15;
}
```

```
.marquee span {
    font-style: italic;
}

@keyframes marquee {
    0% {
        transform: translate3d(var(--move-initial), 0,
0);
    }

    100% {
        transform: translate3d(var(--move-final), 0,
0);
    }
}
```

CSS MarqueeMenu plugin.

When you paste this link in the chrome URL address bar (https://use.typekit.net/zhq0vyf.css) after that you will get this code on the page given below:

```
@import url("https://p.typekit.net/p.css?s=1&k=zhq0vyf
&ht=tk&f=24537.24538.24539.24540.24547.38192.38197.381
98.38199.38200&a=1494256&app=typekit&e=css");

@font-face {
font-family:"sofia-pro";
src:url("https://use.typekit.net/af/0c5f71/00000000000
000003b9b1aa0/27/l?primer=7cdcb44be4a7db8877ffa5c0007b
8dd865b3bbc383831fe2ea177f62257a9191&fvd=n9&v=3")
```

```
format("woff2"),url("https://use.typekit.net/af/0c5f71
/00000000000000003b9b1aa0/27/d?primer=7cdcb44be4a7db88
77ffa5c0007b8dd865b3bbc383831fe2ea177f62257a9191&fvd=n
9&v=3") format("woff"),url("https://use.typekit.net/
af/0c5f71/00000000000000003b9b1aa0/27/a?primer=7cdcb44
be4a7db8877ffa5c0007b8dd865b3bbc383831fe2ea177f62257a9
191&fvd=n9&v=3") format("opentype");
font-display:auto;font-style:normal;font-
weight:900;font-stretch:normal;
}

@font-face {
font-family:"sofia-pro";
src:url("https://use.typekit.net/af/5dd13e/00000000000
000003b9b1a9f/27/l?primer=7cdcb44be4a7db8877ffa5c0007b
8dd865b3bbc383831fe2ea177f62257a9191&fvd=i9&v=3")
format("woff2"),url("https://use.typekit.net/af/5dd13e
/00000000000000003b9b1a9f/27/d?primer=7cdcb44be4a7db88
77ffa5c0007b8dd865b3bbc383831fe2ea177f62257a9191&fvd=i
9&v=3") format("woff"),url("https://use.typekit.net/
af/5dd13e/00000000000000003b9b1a9f/27/a?primer=7cdcb44
be4a7db8877ffa5c0007b8dd865b3bbc383831fe2ea177f62257a9
191&fvd=i9&v=3") format("opentype");
font-display:auto;font-style:italic;font-
weight:900;font-stretch:normal;
}
```

Another example:

```
<!DOCTYPE html>
<html lang="en" class="no-js">

<head>
  <meta charset="UTF-8" />
  <meta name="viewport" content="width=device-width,
initial-scale=1">
  <title>CSS-only Marquee Menu Effect | Codrops</
title>
  <meta name="description" content="A menu with a
css-only marquee hover effect" />
  <meta name="keywords" content="marquee, css,
animation, loop, infinite, hover, menu, navigation" />
```

```
    <meta name="author" content="Codrops" />
    <link rel="shortcut icon" href="favicon.ico">

<style>
*{box-sizing: border-box;margin: 0}body{overflow-x:
hidden;}.content{padding-left: 20px;padding-bottom:
30px}h1{margin: 50px 0 20px;font-family: roboto}
p{margin-bottom: 20px}

.marquee {
  --pos-x: 0;
  width: 100vw;
  display: flex;
  overflow-x: hidden;
}

.marquee__row {
  --translateX: calc(var(--pos-x) * 1px);
  flex-shrink: 0;
  min-width: 100vw;
  display: flex;
  justify-content: space-around;
  transform: translateX(var(--translateX));
}

.marquee--text {
  background-color: #89ff91;
}

.marquee__item--text {
  margin: 0;
  font-size: 65px;
  margin:. 2em. 4em;
}

.marquee--nezuko {
  background-color: pink;
}

.marquee__item--nezuko {
  --height: calc(100px + 50px * ((var(--viewport)
- 375) / 1065));
```

```
  height: var(--height);
  margin: calc(0.1 * var(--height)) calc(0.3 *
var(--height));
}
</style>

</head>

<body>
  <div class="marquee marquee--text" data-speed="25">
    <div class="marquee__row marquee__row--text">
      <p class="marquee__item marquee__item--
text"  data-clone="5">  CSS Marquee Menu Effect / </p>
    </div>
  </div>
</body>

<script>

const marqueeArr = document.querySelectorAll
('.marquee');
marqueeArr.forEach(marquee => {
  const marqueeRow = marquee.querySelector
('.marquee__row');
  const marqueeItem = marqueeRow.querySelector
('.marquee__item');
  const cloneNum = Number(marqueeItem.
getAttributeNode('data-clone').value);
  for (let i = 1; i < cloneNum; i++) {
    const clone = marqueeItem.cloneNode(true);
    marqueeRow.appendChild(clone);
  }
  for (let i = 0; i < 2; i++) {
    const clone = marqueeRow.cloneNode(true);
    marquee.appendChild(clone);
  }

  const marqueeMove = (dir) => {
    const rows = marquee.querySelectorAll
('.marquee__row');
    const rowWidth = rows[0].getBoundingClientRect().
width;
```

```
    let currentX = Number(getComputedStyle(marquee).
getPropertyValue('--pos-x'));
    let newX = 0;
    switch (dir) {
      case 'left':
        newX = currentX?  (currentX - 1) : -rowWidth;
        (newX < (-2 * rowWidth)) && (newX =
-rowWidth);
        break;
      default:
        newX = currentX?  (currentX + 1) : -rowWidth;
        (newX > 0) && (newX = -rowWidth);
    }
    marquee.style.setProperty('--pos-x', newX);
  };

  let speed = Number(marquee.getAttributeNode('data-
speed').value);
  let direction = 'left';
  let marqueeInterval = setInterval(marqueeMove,
speed, direction);
  marquee.onmouseenter = () => {
    clearInterval(marqueeInterval);
  }
  marquee.onmousemove = () => {
    clearInterval(marqueeInterval);
  }
  marquee.onmouseleave = () => {
    clearInterval(marqueeInterval);
    marqueeInterval = setInterval(marqueeMove, speed,
direction);
  }

  let posY = 0;
  const changeDir = () => {
    clearInterval(marqueeInterval);
    let scrollTop = document.documentElement.
scrollTop;
    direction = (scrollTop > posY)?  'right' : 'left';
    marqueeMove(direction);
    marqueeMove(direction);
    marqueeInterval = setInterval(marqueeMove, speed,
direction);
```

```
    posY = scrollTop;
  };
  window.addEventListener('scroll', changeDir);

});

s
</script>

</html>
```

Menu Effect /   CSS Marquee Menu Effect /   CSS Marquee Men

CSS MarqueeMenu plugin.

CHAPTER SUMMARY

In this chapter, we looked at various CSS plugins. It improves the coding functionality by adding these to your plugins.

Appraisal

If you want to be a front end developer, you should have a good quality book, which makes you the developer from scratch to advance. We believe, nowadays the development is on the peek. You might have seen that many biggest companies start their work from scratch by developing new technologies. They all start learning and implementing them in their products. So first, with time, the fundamentals of front end development skills include only some web technologies like HTML, CSS, JavaScript, and many more. You can now extend their functionality by using the third parties packages. Second, the world of technology is constantly changing. With the growing demand of technologies to provide the effective code product, various companies introduce all the new technologies.

To make or look at your website, we use HTML with CSS. Basically, CSS is used to design any page or website.

CSS makes the front end of a website clean and it also creates a great user experience. Without CSS, websites could be less attractive to others. In addition to layout and format, CSS is only responsible for font color and so on.

- CSS helps to keep the informational content of a document separate.

- It also helps us in various ways such as:

  - It is used to avoid duplication.

  - It can take care of maintenance well.

  - It is the same content with different styles for different purposes.

DOI: 10.1201/9781003358060-6

Your website may have thousands of pages that look similar. With CSS, you store style information in common files that everyone can share. When a user displays a web page, the browser loads the style along with the content of the page. When a user prints a webpage, you can provide information about a different style that makes the printed page easy to read.

In general, HTML is used to describe the content of a document, not its style; you use CSS to determine its style, not its content. Of course, there are various exceptions to the rule, and HTML also provides some ways to specify style. For example, in HTML you can use the <b> tag to make text bold, and in its <body> tag you can specify the background color of the page. When you use CSS, you usually avoid using these HTML style features, so all of your document's styling information is in one place.

CSS stands for Cascading Style Sheets. It is a language used to stylize a web page. Also, changing the look and layout of a web page using it is absolutely simple. You can also control how a website appears on different screens of different devices such as mobile phones, desktops, and tablets. You must have a thorough understanding of style sheet language to use it properly.

You can set the font, color, and size for an entire web page or a specific HTML element. However, a single CSS can be redirected to multiple web pages, allowing you to change the look of multiple pages at once. CSS is easy to learn and understand and provides a robust command over viewing an HTML document. Often CSS is combined with markup languages, that is, HTML or XHTML.

HOW CASCADING STYLE SHEETS WORK

Cascading Style Sheet has a completely different approach to a web page style and layout. Every time an HTML document is viewed in a browser, the content comes with style information. Basically, the HTML file contains the content of a page and the style sheet contains information about the style of a page. Therefore, the main purpose of Cascade Style Sheets is to allow elements to appear in the HTML document. And these specified standards guide how content will be executed.

There are various versions of CSS such as:

- CSS1

- CSS2

- CSS3

There are various types of CSS such as:

- Embedded CSS

- Inline CSS

- External CSS

- Import CSS (link tag)

WHY SHOULD SOMEONE LEARN CSS?

- It is must to learn the language for web development.

- Using it you will be able to design unique looking websites.

- Gives you access to almost any item that appears on the screen.

- You can shape any item to your liking.

- You can add effects to great looking web pages.

- You can create animations such as moving objects or shapes using CSS.

- Learning CSS lets you design responsive websites that fit any device size.

- If you learn and understand, you can learn and work with many CSS-based frameworks like Bootstrap.

PROPERTIES OF CSS

- Knowledge of CSS and HTML is essential if anyone wants to pursue a career in web design professionally.

- With the use of CSS, we can control various styles such as text color, font style, spacing between paragraphs, column size and layout, background color and images, layout design, display variations for different screens and device sizes, and many other effects too.

- CSS has control over HTML documents, so it's easy to learn. It is integrated with HTML and XHTML markup languages.

- Understanding of other associated technologies such as Angular, PHP, and JavaScript becomes clearer once we know some of the basics of CSS and HTML.

In CSS, you can use various CSS properties such as:

- CSS Color Codes: It is a color picker, color charts, and code examples all in one page. It includes hex values, RGB, color names, transparency, and opacity.

- CSS @-Rules: There is a list of CSS at-rules. At-rules (or @-rules) define processing rules or values for the stylesheet. They start with an @ followed by their name.

- CSS3 Properties: There is a list of properties introduced in CSS3. It includes animations, flexbox, gradients, multi-column layouts, and more.

- CSS Marquees: To create scrolling text and images in a standards-compliant way, we use CSS Marquees.

- CSS Patterns: CSS gradients are just fading from one color to another.

- You can use gradients to create interesting background patterns.

- CSS Media Types: It is used with media queries. You can apply a separate style depending on the media type that's displaying your web page.

- CSS Media Features: You can use media queries to apply a separate style depending on the media features available in the output device such as phone, tablet, and laptop.

- Floating Menu: You can add a quick code to make a hovering menu.

- CSS Background Color: You can also set the background color of an HTML element.

- CSS Leading: You can also apply the CSS equivalent of leading to your text.

- CSS Align: You can align your elements vertically and horizontally.

- CSS Table Width and Cellpadding: You can use CSS to set the width of your tables with cellpadding.

- CSS Cellspacing: You can modify the space between table cells using CSS.

- CSS Scrollbars: You can use CSS to add scrollbars to an HTML element when its contents become too big.

- CSS Print Version: You can use CSS to apply a separate style to the printed version of your web pages.

Bibliography

1. HTML & CSS – https://www.w3.org/standards/webdesign/htmlcss, accessed on May 16, 2022.
2. CSS – https://developer.mozilla.org/en-US/docs/Web/CSS, accessed on May 16, 2022.
3. CSS Introduction – https://en.wikipedia.org/wiki/CSS, accessed on May 16, 2022.
4. HTML and CSS Basic – https://www3.ntu.edu.sg/home/ehchua/programming/webprogramming/HTML_CSS_Basics.html, accessed on May 16, 2022.
5. CSS History – https://www.w3.org/Style/CSS20/history.html, accessed on May 17, 2022.
6. CSS History – https://www.bu.edu/lernet/artemis/years/2020/projects/FinalPresentations/HTML/historyofcss.html, accessed on May 17, 2022.
7. History overview HTML – https://simplecss.eu/css-history-brief-overview.html, accessed on May 17, 2022.
8. History CSS – https://css-tricks.com/look-back-history-css/, accessed on May 17, 2022.
9. CSS Syntax – https://www.w3schools.com/css/css_syntax.asp, accessed on May 17, 2022.
10. CSS Syntax – https://developer.mozilla.org/en-US/docs/Web/CSS/Syntax, accessed on May 18, 2022.
11. CSS Rules – https://learn.saylor.org/mod/book/view.php?id=36821&chapterid=20180, accessed on May 18, 2022.
12. CSS Selectors = https://developer.mozilla.org/en-US/docs/Web/CSS/CSS_Selectors, accessed on May 18, 2022.
13. CSS Comments – https://developer.mozilla.org/en-US/docs/Web/CSS/Comments#:~:text=A%20CSS%20comment%20is%20used,the%20layout%20of%20a%20document, accessed on May 18, 2022.
14. CSS keywords – https://www.w3.org/wiki/CSS/Properties/color/keywords, accessed on May 18, 2022.
15. CSS – https://www.quackit.com/css/, accessed on May 18, 2022.
16. CSS Properties – https://www.tutorialrepublic.com/css-reference/css3-properties.php, accessed on May 19, 2022.
17. Media queries – https://climbtheladder.com/responsive-web-design-interview-questions/#:~:text=Media%20queries%20are%20a%20key,all%20in%20the%20same%20document, accessed on May 19, 2022.

18. CSS Media Queries – https://www.w3schools.com/css/css_rwd_mediaqueries.asp, accessed on May 19, 2022.
19. Advantages and Disadvantages CSS – https://www.geeksforgeeks.org/advantages-and-disadvantages-of-css/#:~:text=Advantages%20of%20CSS%3A&text=Web%20designers%20needs%20to%20use,web%20site%20and%20maintenance%20time, accessed on May 19, 2022.
20. Awesome PostCSS Plugins – https://www.hongkiat.com/blog/postcss-plugins/, accessed on May 20, 2022.
21. CSS Basics – https://www.cssbasics.com/full.pdf/, accessed on May 20, 2022.
22. Produce Maintainable CSS With Sass – https://openclassrooms.com/en/courses/5625786-produce-maintainable-css-with-sass/, accessed on May 20, 2022.
23. Introduction of CSS – https://techtrendstutor.blogspot.com/p/introduction-of-css-cascading-style.html/, accessed on May 20, 2022.
24. RGB Color – HTML and CSS Guide - https://www.freecodecamp.org/news/rgb-color-html-and-css-guide/, accessed on May 20, 2022.
25. PostCSS – https://github.com/csstools/postcss-preset-env/, accessed on May 21, 2022.
26. Autoprefixer – https://www.npmjs.com/package/autoprefixer/v/5.1.1/, accessed on May 21, 2022.
27. Using Autoprefixer – https://blog.codepen.io/2014/03/28/new-feature-autoprefixer/#docs-nav/, accessed on May 21, 2022.
28. Introduction to PostCSS – https://flaviocopes.com/postcss/#autoprefixer/, accessed on May 21, 2022.
29. PostCSS Nested – https://www.npmjs.com/package/postcss-nested/, accessed on May 21, 2022.
30. PostCSS Plugins – https://github.com/postcss/postcss/blob/main/docs/plugins.md/, accessed on May 22, 2022.
31. Installation – https://github.com/peterramsing/lost/wiki/Installation/, accessed on May 22, 2022.
32. CSS Selectors – https://www.w3schools.com/css/css_selectors.asp, accessed on May 22, 2022.
33. border – https://css-tricks.com/almanac/properties/b/border/, accessed on May 22, 2022.
34. CSS transform – https://www.quackit.com/css/css3/properties/css_transform.cfm/, accessed on May 23, 2022.
35. CSS brightness() Function – https://www.quackit.com/css/functions/css_brightness_function.cfm, accessed on May 23, 2022.
36. CSS – https://www.quackit.com/css/, accessed on May 24, 2022.
37. CSS Properties – https://www.quackit.com/css/properties/, accessed on May 24, 2022.
38. CSS Functions – https://www.quackit.com/css/functions/, accessed on May 24, 2022.
39. CSS Font Properties – https://www.quackit.com/css/tutorial/css_font.cfm, accessed on May 24, 2022.

40. CSS Tutorial – https://www.w3schools.com/css/default.asp accessed on, accessed on May 24, 2022.
41. CSS: Cascading Style Sheets – https://developer.mozilla.org/en-US/docs/Web/CSS, accessed on May 24, 2022.
42. Why to Learn CSS? – https://www.tutorialspoint.com/css/index.htm, accessed on May 24, 2022.
43. Selector – https://en.wikipedia.org/wiki/CSS, accessed on May 24, 2022.
44. CSS Properties and CSS Rules – https://jenkov.com/tutorials/css/css-properties-css-rules.html#:~:text=A%20CSS%20rule%20is%20a,target%20with%20the%20CSS%20rule, accessed on May 25, 2022.
45. CSS - @ Rules – https://www.tutorialspoint.com/css/css_at_rules.htm, accessed on May 25, 2022.
46. Syntax – https://developer.mozilla.org/en-US/docs/Web/CSS/At-rule, accessed on May 25, 2022.
47. CSS rules that will make your life easier – https://www.freecodecamp.org/news/css-rules-to-live-by-962a051e1eb2/, accessed on May 25, 2022.
48. CSS Structure and Rules – https://www.htmlhelp.com/reference/css/structure.html, accessed on May 25, 2022.
49. CSS Position – https://www.javatpoint.com/css-position, accessed on May 25, 2022.
50. CSS Height, Width and Max-width – https://www.w3schools.com/css/css_dimension.asp, accessed on May 25, 2022.
51. CSS Generated Content – https://developer.mozilla.org/en-US/docs/Web/CSS/CSS_Generated_Content#:~:text=CSS%20Generated%20Content%20is%20a,circumstances%20with%20a%20generated%20value, accessed on May 25, 2022.
52. CSS Flexible Box Layout – https://developer.mozilla.org/en-US/docs/Web/CSS/CSS_Flexible_Box_Layout#:~:text=CSS%20Flexible%20Box%20Layout%20is,of%20items%20in%20one%20dimension, accessed on May 25, 2022.

Index